SOCIETY FOR EXPERIMENTAL BIOLOGY

SEMINAR SERIES · 4

ISOLATION OF PLANT GROWTH
SUBSTANCES

ISOLATION
OF PLANT GROWTH
SUBSTANCES

Edited by

J. R. HILLMAN

Department of Botany, University of Glasgow

CAMBRIDGE UNIVERSITY PRESS

CAMBRIDGE

LONDON · NEW YORK · MELBOURNE

Published by the Syndics of the Cambridge University Press
The Pitt Building, Trumpington Street, Cambridge CB2 1RP
Bentley House, 200 Euston Road, London NW1 2DB
32 East 57th Street, New York, NY 10022, USA
296 Beaconsfield Parade, Middle Park, Melbourne 3206, Australia

First published 1978

Printed in Great Britain at the
University Press, Cambridge

Library of Congress Cataloguing in Publication Data
Main entry under title:

Isolation of plant growth substances

(Seminar series – Society for Experimental Biology; 4)
Includes index.
1. Plant hormones – Analysis – Addresses, essays,
lectures. I. Hillman, John R. II. Series:
Society for Experimental Biology (Gt. Brit.) Seminar
series – Society for Experimental Biology; 4.
QK731.I76   582'.03'1   78-1641

ISBN 0 521 21866 7 hard covers
ISBN 0 521 29297 2 paperback

CONTENTS

CONTRIBUTORS

Butcher, D. N. Insecticides and Fungicides Department, Rothamsted Experimental Station, Harpenden, Hertfordshire AL5 5NR, UK.

Crozier, A. Botany Department, The University, Glasgow G12 8QQ, UK.

Gaskin, P. School of Chemistry, The University of Bristol, Bristol BS8 1TS, UK.

Hillman, J. R. Botany Department, The University, Glasgow G12 8QQ, UK.

Horgan, R. Department of Botany amd Microbiology, University College of Wales, Penglais, Aberystwyth SY23 3DA, UK.

McDougall, J. Botany Department, The University, Glasgow G12 8QQ, UK.

MacMillan, J. School of Chemistry, The University of Bristol, Bristol BS8 1TS, UK.

Mousdale, D. M. A. ARC Unit of Developmental Botany, 181A Huntingdon Road, Cambridge CB3 0DY, UK.

Osborne, D. J. ARC Unit of Developmental Botany, 181A Huntingdon Road, Cambridge CB3 0DY, UK.

Powell, R. G. ARC, 160 Great Portland Street, London W1N 6DT, UK.

Reeve, D. R. Botany Department, The University, Glasgow G12 8QQ, UK.

Roberts, J. A. ARC Unit of Developmental Botany, 181A Huntingdon Road, Cambridge CB3 0DY, UK.

Saunders, P. F. Department of Botany and Microbiology, University College of Wales, Penglais, Aberystwyth SY23 3DA, UK.

Self, R. Food Research Institute, Colney Lane, Norwich NOR 70F, UK.

Ward, T. M. ARC Unit of Developmental Botany, 181A Huntingdon Road, Cambridge CB3 0DY, UK.

Wright, M. ARC Unit of Development Botany, 181A Huntingdon Road, Cambridge CB3 0DY, UK.

PREFACE

This book was originally conceived early in 1976 as a result of discussions with the Publications Committee of the Society for Experimental Biology. There was general agreement that a book was urgently required that would bring together the modern methods and techniques involved with the analysis of the known endogenous plant growth substances. The steering committee of the Plant Growth Substances Group, Professor P. F. Wareing, Dr Daphne J. Osborne, Dr D. L. Laidman and Dr J. R. Hillman, then organised a one-day symposium at the SEB Birmingham meeting in early January 1977. All the speakers were invited from centres that have actively encouraged research into plant hormones, and their contributions to that meeting were employed in the production of this volume. The growth substances selected for treatment were restricted to indole acetic acid, the gibberellins, the cytokinins, abscisic acid and ethylene; in future it may be possible to extend this list.

The authors were aware of the difficulties facing the plant physiologist in entering an area of research which concerns an aspect of natural product chemistry. Wherever possible, the articles include criticisms of the various methods, as well as many useful hints so often omitted from materials and methods sections of research papers.

It is hoped that the information presented in this work will find direct and useful application in research and advanced teaching laboratories.

The authors are thanked for their kindness, patience and co-operation. Drs Jack Hannay and John Dale provided staunch encouragement during the whole project and the assistance of Cambridge University Press in overcoming the publishing problems is much appreciated. Professor M. B. Wilkins is also thanked for his guidance and invaluable discussions.

June 1977

J. R. Hillman
Editor for the Society for Experimental Biology

ABBREVIATIONS
AND EQUIVALENT TERMS

α	relative retention; specific rotation
δ^a	nucleophilicity
δ^b	dispersive index
δ^o	polarity
10^{-12} m^3	nanolitre (nl)
10^{-12} m^3 dm^{-3}	nanolitres per litre
ABA	abscisic acid
b. pt	boiling point
BSA	*bis*-trimethylsilylacetamide
BSTFA	*bis*-(trimethylsilyl)trifluoroacetamide
CI	chemical ionisation
cm^3	millilitre (ml)
cpm	counts per minute
DEAE	diethylaminoethyl
dm^3	litre (l)
DMSO	dimethylsulphoxide
dpm	disintegrations per minute
ECD	electron-capture detector
EI	electron-impact
EI–MS	electron-impact mass spectrometry
F_b	background fluorescence level
F_i	'initial' fluorescence level
FI	field ionisation
FID	flame-ionisation detector
FSD	full-scale deflection
GA(s)	gibberellin(s)
GABE(s)	gibberellin benzyl ester(s)
GC–MS	combined gas chromatography–mass spectrometry
GLC	gas–liquid chromatography
GPC	gel permeation chromatography
HFB	heptafluorobutyryl
HFBI	heptafluorobutyryl imidazole
HPLC	high performance (pressure) liquid chromatography
IAA	indole-3-acetic acid
i.d.	internal diameter
IR	infra-red

ABBREVIATIONS AND EQUIVALENT TERMS

M^+	molecular ion
m/e	mass-to-charge ratio
Me	methyl
MF	mass fragmentography (MID, MPM, SICM, SIM)
MID	multiple ion detection (MF, MPM, SICM, SIM)
mm^3	microlitre (μl)
$mm^3\ dm^{-3}$	microlitres per litre
$mol\ dm^{-3}$	moles per litre
MP	mercuric perchlorate in perchloric acid
MPM	multiple peak monitoring
MS	mass spectrometry (see CI, EI and FI)
NMR	nuclear magnetic resonance
OD	optical density
ORD	optical rotatory dispersion
PAW	propan-2-ol:ammonia:water
PC	paper chromatography
PFK	perfluorokerosene
PFP	pentafluoropropyl
ppb	parts per billion
ppm	parts per million
psi	pounds per square inch
PVC	polyvinylchloride
PVP	polyvinylpyrrolidone
RFE	rotary film evaporation
SICM	single-ion (SIM) or selected-ion (MF, MID, MPM, SIM) current monitoring
SIM	single-ion (SICM) or simultaneous-ion (MF, MID, MPM, SICM) monitoring
TFA	trifluoroacetic anhydride
THF	tetrahydrofuran
TIC	total ion current
TLC	thin-layer chromatography
TMSi	trimethylsilyl
UV	ultraviolet

J. McDOUGALL & J. R. HILLMAN

Analysis of indole-3-acetic acid using GC–MS techniques

Introduction

The study of phototropism in coleoptiles of etiolated monocotyledonous seedlings was instrumental in establishing the concept of growth hormones in higher plants. Observations by Charles & Francis Darwin (1880), Rothert (1894), Boysen-Jensen (1910), Paál (1919) and Went (1928), together with the results published by Söding (1923, 1925), led to the generally accepted theory that the tip produces or secretes indole-3-acetic acid (IAA) which, in turn, regulates normal elongation growth in the physiologically more mature cells in the part below. What is more, the curvature seen to result from irradiating the coleoptile from one side could be explained in terms of an unequal distribution of IAA on the lighted and shaded sides, causing growth towards the light source (see Thimann, 1972). This interpretation, founded on the redistribution of IAA brought about by an environmental factor, was also applied to the action of gravity in bringing about geotropic curvature (Navez & Robinson, 1933; Dijkman, 1934; Dolk, 1936).

Based mainly on crude analyses of extracts and especially investigations of the effects, transport and metabolism of the *exogenous* compound, IAA has come to occupy a pivotal role in the regulation of many aspects of growth and differentiation in a wide range of species. Certainly, exogenous IAA affects the pattern of a large number of developmental phenomena, and this is taken as an indication that IAA could act in concert with other substances known to influence growth. Any critical assessment of the significance of a putative hormone in the life of a plant, however, must take many factors into consideration; not least of these is the unequivocal demonstration of the presence of that compound in the plant. Information about amounts produced in a particular tissue, cell or cell component is also necessary in revealing any correlative role. Many more criteria are demanded but these are beyond the scope of this article.

The physiologist then, must have the ability and resources to isolate, identify and measure the hormone in question. In this article we discuss various procedures that can be used in achieving this aim in respect of IAA, and our recommendations are made on the basis of experience of the techniques available at the present time. We have judged combined GC–MS to be the most desirable and convenient form of confirming the presence of IAA in extracts.

With access to sophisticated analytical instruments, harsh judgement can readily be passed on the results and interpretations of earlier experimenters, but the true eminence of the pioneer studies is more than apparent when seen in the context of the times and the limited facilities at their disposal. Nevertheless, if the aim of the experiment is to investigate IAA *per se*, then little of value can now arise from the continued use of outmoded techniques. There is no excuse for adding to the wealth of existing literature by describing what are essentially 'IAA-like' chemical properties, as occurs when bioassay and separative procedures alone are employed to identify IAA.

There are many compounds which have similar biological activity to IAA in certain assay systems, and the term 'auxin' is generally applied to such substances. Some auxins are produced naturally whereas others are produced artificially. This article, however, is restricted to the isolation and measurement of *free* IAA from plant tissues. It must be stressed that the following methods require careful adaptation to suit the equipment available, type of tissue or extract, and aims of the experiment.

The compound

IAA is a compound which has been given a selection of names (Fig. 1). When pure it consists of leaflets of a white crystalline powder; any alteration is evidenced by the appearance of a reddish colour and the development of a distinctive indolic smell. The presence of oxygen and light assist in breakdown. It is freely soluble in methanol and ethanol,

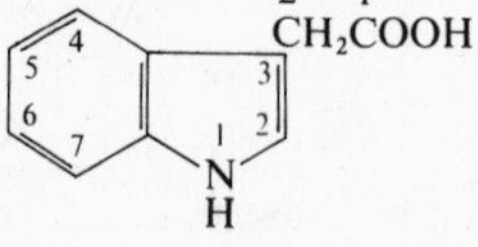

Fig. 1. The formula of IAA (auxin, indolyl acetic acid).

soluble in acetone, ethyl acetate and diethyl ether, and sparingly soluble in water and chloroform.

Procedures

The whole procedure recommended for heavily pigmented samples (Table 1) can be divided into six stages: (1) Extraction, (2) Preliminary purification, (3) Column chromatography, (4) Thin-layer chromatography (TLC), (5) Gas–liquid chromatography (GLC), (6) Mass spectrometry (MS). Stages (5) and (6) are usually, but not invariably, combined (GC–MS).

The aim of this processing is to increase the concentration of IAA in the sample by decreasing the amount of unwanted materials, thereby providing a sample of low dry weight. This is important because dry-weight contamination must be reduced to a minimum prior to GC–MS analysis.

Table 1. *Complete basic purification and identification procedure*

(1) Extraction	Extract with methanol.
(2) Preliminary purification	Filter and obtain dry weight of residue. Add internal standard to extract. Reduce by RFE to aqueous condition. Filter solution including washings of flask. Acidify to pH 3.0 using 2 mol dm^{-3} HCl. Partition 3 times against diethyl ether. Collect ether layers and discard aqueous phase. Reduce by RFE and dissolve residue in methanol.
(3) Column chromatography	Apply to DEAE-cellulose column (15 g). Wash column with water and discard. Elute column with 0.05 mol dm^{-3} sodium sulphate. Acid partition the eluate against ether. Reduce ether by RFE and dissolve residue in methanol. Apply to PVP column (20 g). Elute column with 0.1 mol dm^{-3} phosphate buffer. Acid partition the eluate against ether. Reduce ether by RFE and dissolve residue in methanol.
(4) TLC	Apply to TLC plate and develop. Elute IAA R_f with methanol. Reduce in volume with nitrogen, take up in ether. Transfer solution to fresh vial, remove ether in nitrogen and calculate dry weight.
(5) GC and (6) MS	Derivatise for 3 h at 60 °C using BSA. Carry out radio-GLC and combined GC–MS.

RFE = rotary film evaporation; BSA = *bis*-trimethylsilylacetamide.

Extraction

Fresh tissues are to be preferred and suitable harvesting methods should be employed to avoid contamination. Freeze-drying has been stated to be the most appropriate method of storing plant material (Audus, 1972) but, if possible, samples should be processed immediately and the extract taken to the derivatisation stage prior to GC–MS.

Five criteria must be borne in mind when selecting a suitable method of extraction.

(1) All of the free IAA should be extracted from the tissue with minimal contamination by unwanted materials, organisms or tissues.

(2) No synthesis, conversion or degradation of the extracted IAA should occur during the extraction process.

(3) Only *free* IAA should be extracted such that a separate assessment can be made of the bound fraction in the residue.

(4) The IAA in the extract must not arise as an artifact, e.g. microbial activity. Segments have been frequently used in physiological experiments and it has yet to be satisfactorily demonstrated that the IAA measurements made have not reflected cut-surface effects.

Collections can be made in agar for subsequent GC–MS analysis (Greenwood *et al.*, 1972; White, Medlow, Hillman & Wilkins, 1975) but a considerable number of individual diffusions have to be employed. It has been argued that the so-called 'diffusible' IAA collected by this process may represent the physiologically active component (see Audus, 1972) though care is needed in providing interpretations, e.g. the contribution of the cut surface, development of a secondary physiological tip, secretion problems, re-entry of IAA from agar to tissues, etc. We have noted complications in the *Zea* seedling root system: although IAA has been found by direct extraction of root caps (Rivier & Pilet, 1974), diffusions of 26 000 root caps into specially purified agar could not reveal the presence of IAA using similar GC–MS identification procedures (Hillman, unpublished).

Ion-exchange beads, buffers and water may be used as extraction media but careful monitoring of the five points mentioned above is required. Many of the extraction and subsequent purification problems can, however, be overcome by the use of xylem and phloem saps, and, in the future, techniques such as differential centrifugation may have an important part to play in the preparation of extracts.

The most typical method of extraction is with volatile organic solvents and there have been some early reports on their relative efficacy (Larsen, 1955; Nitsch, 1956; Bentley, 1961; Pilet, 1961). Whitmore & Zahner (1964) provided some evidence that, as a result of polyphenol activity, IAA synthesis could occur during ether extraction of *Pinus* phloem and cambial tissue.

Methanol would appear to be the most efficient solvent we have tested; it decolourises the tissues, indicating plastid breakdown; it has denaturing properties allowing rapid entry; and it appears to give a higher yield of IAA from tissues than acetone, ethanol, diethyl ether, chloroform and ethyl acetate. Atsumi, Kuraishi & Hayashi (1976) claimed on the basis of simple and non-definitive purification methods that transamination of tryptophan to indole pyruvic acid, followed by decarboxylation to IAA, could occur during acidified methanolic extraction of plant tissue. Were this to occur then the final estimations of IAA in plant extracts could be inaccurate and further investigation of this aspect is merited.

If methanol is selected as the extractant then it must be redistilled. Either absolute or 80% methanol is used to prepare the extract with a ratio of 1 g fresh weight of tissue to a minimum of 5–10 cm^3 of solvent. To date, there is no convenient method whereby the efficiency of extraction can be accurately monitored: this would require a study of the removal of labelled endogenous IAA. The arbiter of extraction is therefore taken to be the complete removal of chlorophyll and other pigments. Enhancement of the extraction rate is usually obtained by both occasional agitation of the solution and the use of fresh solvent. Even for the extraction of large quantities of material the extraction time should not exceed 24 h, and in the case of small quantities of tissues it is possible to achieve maximum yield with just 1–2 h of extraction. No difference has been found in yield between methanolic extractions at 20–25 and 4 °C (Hillman, unpublished).

Homogenisation and crushing at low temperatures can speed up the extraction phase but must be viewed with caution until definitive experiments can ascertain whether or not the levels of free IAA are affected by enzyme action.

Preliminary purification

Following the extraction phase and before identification is possible using mass spectrometry, it is necessary to purify the sample. During purification the sample volume is reduced to approximately 5–20 mm^3 and unwanted compounds are eliminated to provide a sample of low dry weight.

Preliminary purification using solvent partitioning is a rapid first purification stage that involves minimal loss of any IAA present in the extract. In fact, Kögl used this method in his early studies (Kögl, Haagen Smit & Erxleben, 1934 a, b).

After extraction using methanol the solution is either filtered using

Büchner filtration with proper solvent washing of the filtered residue and apparatus, or centrifuged to remove coarse cell debris. An estimation of the dry weight of the extracted tissue can now be made, whilst appreciating that small quantities of material have been removed in the solvent.

The internal standard may be added either to the filtrate, or earlier when the methanol is brought into contact with the tissues. Internal standards are used to monitor losses occurring during the isolation procedures and are absolutely necessary for any quantitative studies. This aspect is dealt with in the section 'The use of standards and controls', below.

After filtration, the organic solvent can be removed from the extract by rotary film evaporation (RFE) at 30–36 °C, leaving an aqueous residue containing the endogenous IAA. An indication of when the aqueous stage is reached is seen by the marked drop in the rate of evaporation losses from the extract, and by this time green pigmented samples turn a dark colour and become highly viscous. This aqueous fraction is then immediately acidified and partioned against diethyl ether. Once in ether, the total sample volume containing IAA can be easily reduced to a convenient quantity, usually 1–5 cm³, ready for chromatography.

However, on removal of the organic solvent from the extract a considerable amount of solid debris can often be noted in the aqueous phase and on acidic–ether partitioning this debris will dissolve in the ether. By the simple expedient of a filtration step to remove debris from the aqueous residue, it is possible to reduce the dry weight by up to 80%. The aqueous extract therefore is filtered through Whatman No. 3 filter paper using a Büchner apparatus, taking care to rinse the flask thoroughly with three aliquots of 50 cm³ water, which are also filtered. The combined aqueous extract is now ready for acidification to pH 3.0. HCl of low normality is used for acidification to reduce the possibility of localised hydrolysis occurring in the solution on addition of the acid.

Thorough partitioning against equal volumes of diethyl ether should be carried out three times and any IAA present in the extract will pass into the ether layers. The ethereal fractions are combined and reduced in volume to a residue which is taken up in methanol in preparation for the column chromatography stages.

Some investigators use an alkali–ether partitioning stage during preliminary purification, but in the preparation of *Phaseolus* extracts this would appear to be of little effect in reducing sample dry weight.

By the use of radioactive IAA as an internal standard, it is possible to monitor the percentage recovery of IAA at each step in the preliminary purification. At each stage of the extraction, aliquots are taken and the percentage recovery of isotope calculated. The filtration of the aqueous phase and the acidic–ether partitioning have individual recoveries of isotope in the range of 90–93%, with an overall recovery for the whole of the preliminary purification process of 83–85%. Similarly, Mann & Jaworski (1970) have reported recoveries of 92–93% for this stage. It should be noted, however, that they included a backwash of the ether extract with 5 cm^3 of 1 mmol dm^{-3} phosphoric acid.

Column chromatography

Following preliminary purification, extracts can be purified further by column chromatography before TLC and GLC. Most column systems have a greater loading capacity than TLC and GLC.

Ion-exchange resins, notably diethylaminoethyl(DEAE)-cellulose, are particularly useful as a first column purification of extracts (Burnett, Audus & Zinmeister, 1965; Elliott & Greenwood, 1974; Bridges, Hillman & Wilkins, 1973; White et al., 1975; Robertson, Hillman & Berrie, 1976). This column system appears to be an excellent way of eliminating pigments from extracts, with the various pigment fractions being left behind as bands on the column. It also has the added advantage of being fast-running under gravity feeding.

The usual procedure is as follows: 15 g of DE 1 or DE 23 are pre-equilibrated by initially washing in 0.5 mol dm^{-3} HCl, draining and rinsing with water. The DEAE-cellulose is then washed in 0.5 mol dm^{-3} NaOH and then continuously washed in distilled water until a final pH of 7.5 or less is reached. After applying the extract, 300 cm^3 of distilled water are passed through the column to remove unbound material and then discarded. 350 cm^3 of 0.05 mol dm^{-3} sodium sulphate solution are then used to elute the ionically bound substances including the IAA, the recovery of IAA being greater than 90%. The sodium sulphate used to elute the column is acid-partitioned against ether in the usual manner, with the combined ether layers being reduced to a residue which is then taken up in methanol for further chromatography.

Bandurski & Schulze (1974) also used DEAE-cellulose as a preliminary column system but eluted IAA from the column by gradually increasing the concentration of acetic acid in a chloroform–methanol solvent mixture.

Further purification of the sample may be necessary and polyvinyl-pyrrolidone (PVP) is a convenient second column chromatography system, based on liquid–solid adsorption. Glenn, Kuo, Durley & Pharis (1972) used PVP to separate plant hormones in extracts, with 0.1 mol dm^{-3} phosphate buffer as the liquid phase; at a buffer pH of 8.0 IAA was noted to elute between 160 to 290 cm^3. In contrast, using a 20 g column of PVP (Polyclar

AT) and 0.1 mol dm^{-3} phosphate buffer at pH 8.0, we find that IAA elutes between 90 to 170 cm^3 with a recovery of [^{14}C]IAA that exceeds 90%. In practice, chromatography using PVP columns gives a *ca* 50% reduction in sample dry weight. It is efficient at removing pigment from extracts but has a lower loading capacity than DEAE cellulose.

Before use, Polyclar AT must be washed to remove the fine particles. Thus the PVP is dispersed in an excess of distilled water; after standing for 15 min, the excess water is decanted off and the process is repeated a further two times using distilled water and thrice using the buffer.

Once the IAA has been eluted satisfactorily, then the buffer is acid-partitioned against ether and once again the ether is reduced to a residue which is taken up in methanol.

De Yoe & Zaerr (1976) have used silica gel columns with a 1% solution of HCl in 95% ethanol as the eluant. They found that IAA in extracts of Douglas fir eluted between 90 and 170 cm^3.

Partition column chromatography appears to have been little used in purifying plant extracts. Although no plant material was involved, Raj & Hutzinger (1970) separated various indole acids on a column of Sephadex G-25 with a benzene:dioxane:water (1:1:1: v:v:v) mixture.

Various Sephadex gels have been used to purify plant extracts containing IAA. Steen & Eliasson (1969) identified IAA by UV detection in extracts of *Picea abies* after chromatography of the extract on Sephadex LH 20, with ethanol as the solvent. Moreover, Bandurski & Schulze (1974) achieved an 82% recovery of IAA from Sephadex LH 20 eluted with 50% ethanol. Anderson (1968) separated various indoles, including IAA, from cultures of *Claviceps*, using ammonium acetate as a solvent for a column of Sephadex G-10.

Recently, Reeve & Crozier (1976) have utilised gel permeation chromatography to purify various hormones, including IAA, from plant extracts. Separation in this instance occurs on the basis of decreasing molecular size.

Analytical high pressure (performance) liquid chromatography (HPLC) has been used to detect IAA in extracts, with detection being by either UV absorbance or fluorescence, together with the appropriate retention time. Bausher & Cooper (1976) utilised HPLC with a UV-absorbance detector to reveal IAA in xylem sap of *Citrus* rootstocks. Similarly, Durley & Kannangara (1976) detected IAA in extracts of *Sorghum* leaves. An UV-fluorescence detector was used to identify IAA in samples from wheat, soybean, pinto bean, cotton and pea plants.

Preparative HPLC has been used by Brenner *et al.* (1976) to separate

IAA and ABA from soybean, with recoveries of 90%, prior to identification by GC–MS and quantification by GLC.

Thin-layer chromatography

After column chromatography, a TLC stage is frequently used before preparing samples for GLC or GC–MS. As Sagi (1969) has demonstrated, if silica gel is used then the recovery of the IAA from the TLC plate must be completed within 2 h or breakdown of the IAA may occur. This effect, however, is reduced when cellulose is used as the solid support. In addition, we have found that breakdown of IAA may occur on the TLC plate prior to development, a problem which is exacerbated by time and light.

The plant extract is usually applied to the TLC plate as a streak, allowing room either side for two lateral spots of authentic IAA; there are dozens of solvent systems suitable for TLC of IAA (see e.g. Stahl & Kaldewey, 1961; Kaldewey & Stahl, 1964), with solid support of either cellulose or silica gel. Of 18 solvent systems we have tested, 5 give efficient separation of IAA from other components of *Phaseolus* extracts (Table 2).

The methyl acetate:propan-2-ol:ammonia system appears to be one of the best systems; propan-2-ol:ammonia:water suffers from poor resolution between a number of closely related indoles. Any variations that occur in R_f values can be readily eliminated by development under standard conditions of darkness at a constant temperature in a still atmosphere, preferably of nitrogen.

Various methods of detecting IAA on TLC plates are available. Ehrlich's reagent, which is a solution of *p*-dimethylaminobenzaldehyde in HCl will

Table 2. *TLC systems for IAA using silica gel plates*

Solvent system	R_f at 15 °C	[^{14}C]IAA recovery with methanol (%)
Propan-2-ol:35% ammonia:water (100:5:10: v:v:v)	0.24–0.34	50
8% Sodium chloride solution	0.44–0.62	22.6
Methyl acetate:propan-2-ol:25% ammonia (45:35:2: v:v:v)	0.32–0.42	74.5
Chloroform:methanol:35% ammonia (80:25:0.1: v:v:v)	0.01–0.07	67.7
Propan-2-ol:35% ammonia:water (10:1:1: v:v:v)	0.24–0.31	71.2

react with various indoles, including IAA, to give a deep pink colour (see Thimann, 1972). Many variations of the Ehrlich reaction test and other destructive colour reaction tests are described by Bentley (1961). Tirimana & Geevaratne (1972) described a vanillin–sulphuric acid reagent which detected IAA on TLC plates to a limit of sensitivity of 12.5 ng; indole compounds substituted in the 3-position reacted to give a violet colour.

The most convenient method of detection available is probably that based on absorbance of UV light at 254 nm, where the solid support contains an UV fluorescent agent. Such plates are commercially available (e.g. Machery-Nagel), as are many types of suitable UV chromatogram scanners.

Recovery of the IAA from the developed TLC plates is accomplished by eluting the appropriate zone of the solid support with methanol. Sagi (1969) used methanol and Salper reagent (1 cm^3 0.5 mol dm^{-3} $FeCl_3$ in 50 cm^3 35% $HClO_4$) and reported 78% recovery. When silica gel containing an UV agent is used, then some contamination of the extract will arise with methanol elution. In order to avoid this complication it is essential to eliminate the agent by reducing the methanol to dryness and then redissolving the residue in diethyl ether.

Gas–liquid chromatography

Gas–liquid chromatography (GLC) has been extensively employed in the analysis of plant extracts for IAA. It provides a valuable separation stage for low amounts of the compound and should be viewed in this context rather than as a form of identification itself.

Three types of detector have been employed in GLC of IAA. The flame-ionisation detector (FID) has been the most widely used with sensitivity limits approaching 1 ng for almost all organic compounds. A more sensitive detector is found in the electron-capture detector (ECD), but it is more restricted in usage simply because it responds to substances that readily capture electrons. In the case of IAA detection, therefore, it is usual to convert any IAA present in the sample to a halogenated derivative. The most selective detector is the mass spectrometer (MS).

In view of its relatively high melting point, IAA must be converted to a more volatile derivative before GLC analysis. The choice of derivative depends on the nature of the GLC detector.

A well-known derivative is the methyl ester of IAA (Me-IAA) produced either by boron trifluoride and methanol catalysis or by diazomethane such that methylation occurs at the carboxyl group. This ester was studied in the identification of IAA in *Zea* (Powell, 1964) and *Nicotiana* (Bayer, 1969); in both instances identification relied on GLC co-chromatography and spectrophotofluorimetry. Me-IAA was also employed in identifying IAA by GC–MS in Douglas fir (De Yoe & Zaerr, 1976) and *Zea* roots (Elliott & Greenwood, 1974).

Analyses of IAA and other substituted indoles by GLC of methyl esters have also been carried out from sources other than plant material (Dedio & Zalik, 1966; Grunwald, Vendrell & Stowe, 1967; Grunwald & Lockard, 1970).

Fluorinated derivatives of Me-IAA have proved suitable for ECD analysis with a threshold of sensitivity of approximately 100 pg; fluorination occurs at the amino group of IAA. Seeley & Powell (1974) utilised both heptafluorobutyryl (HFB) and trifluoroacetyl (TFA) derivatives of Me-IAA during the analysis of IAA in apple seeds, and they found the TFA to be more suitable. Rivier & Pilet (1974) used the HFB derivative of Me-IAA to identify IAA in *Zea* roots, whereas Hopping & Bukovac (1975) produced the TFA ester of Me-IAA in their GC–MS identification of IAA in *Prunus* fruits. TFA derivatisation of substituted indoles and their analysis by GLC and ECD have also been reported by Brook, Biggs, St John & Anthony (1967).

An attempt to increase the limits of sensitivity of ECD detection of IAA was described by Bittner & Even-Chen (1975). Synthesis of the trichloroethyl ester of IAA improved the limit of response of their ECD to 20 pg, but application of this method to plant extracts was unsuccessful because of the presence of interfering substances.

Trimethylsilyl (TMSi) esters of IAA have been successfully applied in the detection of IAA by FID. This ester is formed by the reaction of IAA with a silylation agent, commonly *bis*-trimethylsilylacetamide (BSA) or *bis*-(trimethylsilyl) trifluoroacetamide (BSTFA); silylation occurs at both the carboxyl and amino group. Bandurski & Schulze (1974) used TMSi-IAA in the GLC and GC–MS identification of IAA in *Avena* and *Zea*. Similarly, Shindy & Smith (1975) used this derivative and methodology to demonstrate IAA in cotton ovules. TMSi-IAA has been used extensively in GC–MS analyses of *Zea* roots (Bridges *et al.*, 1973), *Phaseolus vulgaris* (White *et al.*, 1975; Hillman, Math & Medlow, 1977), *Ricinus communis* (Hall & Medlow, 1974), and *Lactuca* fruits (Robertson *et al.*, 1976).

The choice of derivative relates to the type of detector and convenience of preparation. We find the production of the TMSi derivative to be a relatively simple method of esterification, resulting in a stable product. It has the advantage over the methyl ester in that Me-IAA has been shown to be present in plant tissue (Takahashi *et al.*, 1975), hence IAA estimations based on methylated IAA would fail to allow for a contribution from the endogenous component.

Recommended derivatisation procedures. Preparation of the *bis*-TMSi-IAA derivative is achieved by dissolving a thoroughly dried extract in either BSA or BSTFA and reacting for a suitable time. Silylation of IAA occurs more rapidly in BSTFA than BSA. Bandurski & Schulze (1974) reported that, in the presence of pyridine, BSTFA will silylate both the carboxyl and nitrogen group within 15 min at 50 °C. In the case of BSA as the silylation agent, double derivatives of IAA are noted (see Bosin, Buckpitt & MaicKell, 1974). If IAA is reacted with excess BSA for 1 h at 60 °C then two derivatives are formed; analysis of the mass spectrum of these compounds shows that one is the *bis*-TMSi derivative and the other the monoderivative where silylation is confined to the carboxyl group. Complete silylation occurs after 3 h at 60 °C (Fig. 2).

The main problem encountered in preparing TMSi derivatives is incomplete drying of the extract; even trace amounts of water will lead to solidification of the reaction mixture and destruction of the derivative.

Formation of the methyl ester of IAA is best accomplished by reacting IAA with an ethereal solution of diazomethane. Following esterification, the excess reagent is removed by a stream of nitrogen gas and the residue is dissolved in a suitable volume of solvent. Precautions must be taken, however, to exclude endogenous Me-IAA from the sample.

Me-IAA may be further converted to a fluorinated derivative. According to Bertilsson & Palmér (1972) the HFB derivative, resulting from reaction of the sample with heptafluorobutyryl imidazole (HFBI) for 1 h at 85 °C, is ready for analysis after taking up in *n*-hexane. Me-IAA can also be converted to the trifluoroacetic anhydride (TFA) derivative by reacting in TFA and sodium sulphate for 1 h. Therefore the excess TFA is removed and the sample dissolved in ethyl acetate (Hopping & Bukovac, 1975).

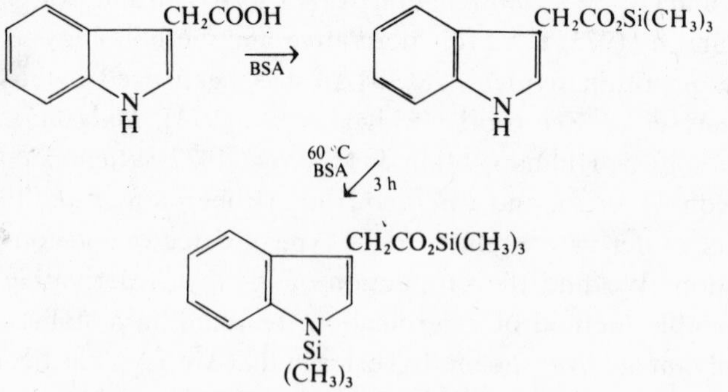

Fig. 2. Derivatisation of IAA using BSA.

Table 3. *Liquid stationary phases and oven temperatures for the GLC analysis of IAA*

Ester	Solvent	Oven temperature (°C)	Column	Plant source	Detector
Me-IAA	Acetonitrile or ethanol	230	7% Vermasid	Maize	FID
Me-IAA	Acetone	190	3% SE-30	*Nicotiana*	FID
Me-IAA	Methanol	200	2.5% Hi-Eff 8-BP	Douglas fir	FID
TFA-Me-IAA	Hexane	140	2% SE-30	Apple	ECD
HFB-Me-IAA	Hexane	140	2% SE-30	Apple	ECD
HFB-Me-IAA	Ethyl acetate	140	3% SE-30 3% DC-200	Cherry	MS
TMSi-IAA	BSA	200–220	2% SE-33	Lettuce	MS
TMSi-IAA	BSTFA and pyridine	155	5% SP-2401	Oat and maize	FID
TMSi-IAA	BSA	100–250	3% QF-1	Cotton	MS

FID = flame-ionisation detector; ECD = electron-capture detector; MS = mass spectrometer.

Columns. The choice of stationary phase and oven temperature obviously depends on the nature of the derivative and extract. Utilising the TMSi derivative with *Phaseolus* extracts we have found that the three gas chromatographically equivalent non-polar stationary phases OV-1, OV 101 and SE-30 at 3–5% concentration on Gas Chrom Q and Chromosorb W will give excellent resolution between IAA and other components in the extracts at 200 °C with a detection limit in practice of 10–20 ng by FID with excess silylation reagent as the solvent.

Other liquid stationary phases that have been utilised in GLC analysis of IAA are shown in Table 3.

Mass spectrometry

Mass spectrometry (MS) is arguably the most powerful method available for the rapid identification of components of biological origin. It is possible to identify a compound by virtue of its molecular weight and/or characteristic fragmentation pattern. Moreover, direct measurement or quantification of the compound is possible by monitoring one or more of the fragment peaks.

Biological samples can be introduced into the MS by direct probe, but this method has the prerequisite that the sample must be of very high purity. More impure samples are commonly purified prior to entry into the MS by first chromatographing the sample on a GC which is coupled directly to the MS. This method of combined GC–MS allows the sample to be separated into individual peaks, each of which can be analysed by the MS. The main problem in combining these instruments comes from transferring the GC effluent at a positive pressure to the MS which has a normal operating vacuum of 10^{-5}–10^{-7} torr. Typically, the interface consists of a molecular separator.

There are two main types of separator used for the analysis of IAA – membrane and jet. In practice a distinct amount of molecular selectivity is achieved with the membrane separator and sample of *ca* 7 mm^3 can be injected into the GC. The jet separator does not have such a degree of selectivity and the samples have to be considerably purer; moreover, injection volumes of only 1 mm^3 or less are required for the GC.

Analysis of a compound by GC–MS can be made by either or both of the following: (1) obtaining a full scan giving the detailed fragmentation pattern of the compound (this inevitably includes some interfering ions); and (ii) single (SIM) or multiple (MPM, MIM, MID, MF, SICM) ion monitoring where prominent ions of the compound and GC retention values are monitored. If only a single ion is monitored then a full scan should be obtained at the appropriate retention time, otherwise the MS can only be considered as a selective detector for the GC.

MS of IAA and its derivatives

(i) *Identification.* When introduced into the MS, indole has a mass spectrum after ionisation by electron impact (EI) consisting of the molecular ion (M^+) as the base ion (i.e. most prominent ion) with very little fragmentation occurring. Introduction of alkyl substituents into the 3-position of the indole system results in an α,β side-chain cleavage with the resulting ion as the base peak. In the case of IAA this ion has a m/e value of 130 (Fig. 3), but its exact nature is unclear. Two structures have been proposed viz. the protonated methyl-3-indole cation or the quinolinium cation. Further fragmentation of this ion occurs by the loss of HCN resulting in an ion of m/e 103, and the subsequent loss of acetylene units giving rise to ions of m/e 77 and m/e 51 (Jamieson & Hutzinger, 1970). This fragmentation pattern is similar for most 3-substituted indoles, and the ions all occur in a characteristic relative abundance (Table 4).

The behaviour of derivatised IAA on EI–MS is similar to that of IAA, with obvious changes in the m/e values of certain fragment ions to allow for addition of the derivative complex.

Typically, for SIM of TMSi-IAA the MS is adjusted to focus only on those ions occurring at m/e 202, and 1 ng should be readily detectable. For much lower quantities, detection is possible but it is important that

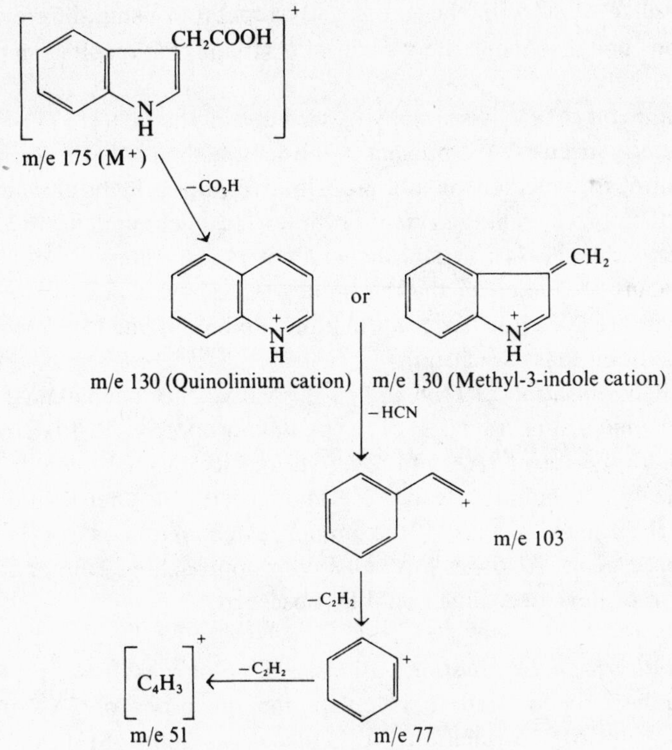

Fig. 3. MS fragmentation pattern of IAA.

careful control is maintained over sample variation and adsorption in the system. Thus interspersed solvent blanks must be used to prevent any memory effects. At the outset, standard TMSi-IAA is injected into the GC and the MS tuned to detect the m/e 202 ion. After injecting the reagent blank, the silylated plant extracts should be examined under identical conditions. The presence of a peak at the correct retention time indicates

Table 4. *Major ions produced in EI-MS of IAA and its derivatives*

Compound	m/e		
	Molecular ion	Base ion	Others
IAA	175	130	103, 77, 51
bis-TMSi-IAA	319	202	304, 73
HFB-Me-IAA	385	326	129, 69
TFA-Me-IAA	285	226	129, 69
Me-IAA	189	130	103, 77, 51

the presence of IAA in the extract. By repetition using different GLC conditions and also using the weaker m/e 319 ion, the results can then be confirmed.

Simultaneous MPM is normally carried out on the m/e 319 and m/e 202 ions. If derivatised IAA is present in the sample then only those channels representing the selected ions show positive responses in the characteristic ratio at the correct relative retention time. This technique has been used by Rivier & Pilet (1974), Takahashi *et al.* (1975) and Hillman *et al.* (1977).

Full scans are taken in those samples that show a positive response in SIM at m/e 202. Further confirmation can be obtained by obtaining a high resolution mass spectrum.

Although ionisation of IAA and its derivatives has been almost totally by EI, chemical ionisation (CI) of the pentafluoropropyl (PFP) derivatives of the methyl ester of IAA and other various amines has been described by Miyazaki, Hashimoto, Iwanga & Kubodera (1974). When ammonia was used as the reagent gas, all of the amines tested yield ions of $(M+NH_4)^+$ as the base peak. Also, an MS sensitivity approaching 1 pg of the PFP derivative of nor-adrenaline could be observed.

(ii) *Quantitative determinations.* Bioassays, colour tests, GC peak areas and MS have been used to calculate the quantities of IAA in plant extracts. For MS estimations it is usual to apply SIM and MPM procedures.

With SIM a calibration curve is constructed plotting the amount of standard IAA injected as a function of peak area or height and a linear response is usually, but not invariably (Robertson *et al.*, 1976), noted. It is recommended also that these injections be repeated in the presence of the plant extract to be tested, as the higher and lower limits of the linear portion may be affected if relatively impure samples are injected into the GC. Thereafter a reagent blank, sample and blank injection sequence is carried out, followed by a further set of standards. It is then possible to calculate peak areas at the correct retention time by comparison with the calibration curve. Quantitative estimations using MPM are carried out in the same manner.

Older techniques of measuring the heights of the base ion in repeat full scans (Bridges *et al.*, 1973; Hall & Medlow, 1974) as well as isotope-dilution (Bandurski & Schulze, 1974) have been used for IAA estimations.

The use of standards and controls

A solvent control carried out in tandem with the extract is essential in any experiment involving extraction and purification steps. Therefore, a volume of deionised water equivalent to that of the tissue is processed to test whether or not IAA is added as a result of cross-contamination.

Internal standards, required to check losses during purification, can be labelled with either a radioactive or a stable isotope, and must have similar chromatographic properties to the compound of interest without affecting estimations and identification.

Mann & Jaworski (1970) demonstrated that the losses incurred during the purification of IAA extracts can be estimated by the addition of radioactive IAA to the initial extract. In the main, there are three types of radioactive IAA commercially available: $1'$-[^{14}C]IAA, $2'$-[^{14}C]IAA and 5-[^{3}H]IAA. It is an established practice to calculate the percentage recovery of isotope by liquid scintillation spectrometry of a small aliquot from the final sample prior to GC–MS. This method of estimating the efficiency of purification reveals only the recovery of the radioactivity, be it ^{14}C or ^{3}H; it does not give an accurate picture of the recovery of the labelled IAA *per se*. This problem can be overcome by chromatographing the sample and analysing the distribution of the radioactivity with a suitable detector. Two such procedures may be undertaken.

One method is to remove an aliquot of the purified sample before derivatisation and carry out TLC. Radioactivity of the sample can then be determined using a TLC-radiochromatogram scanner. Indeed, samples can be taken at various stages of purification to determine the recovery of the isotope. Assuming that the R_f zones of the standard and extract are identical it is possible to calculate the percentage purity of recovered IAA and hence the actual amount of radioactive IAA.

A better method is to determine the content of radioactive IAA after derivatisation using gas–liquid radiochromatography (radio-GLC). This involves connection of the GLC to a radioactivity monitor by means of a splitter between the GLC detector and the column. In this mode a portion of the effluent from the column can be directed for combustion and subsequent radioactivity assessment in a proportional flow counter. The peak of radioactivity in the sample having an identical retention value to that of authentic IAA is compared with a set of reference standards to calculate the actual amount of recovered radioactive IAA. This process normally involves estimation and comparison of peak areas. In Fig. 4 (*a*) refers to a sample which was primed with internal standard before purification and (*b*) refers to the [^{14}C]IAA standard alone. The main peak in the sample A had a similar retention value to that of the standard and

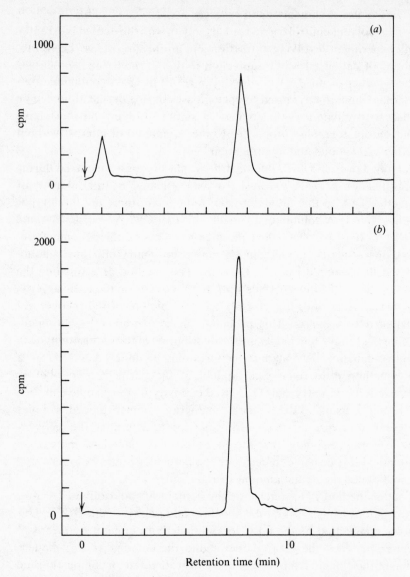

Fig. 4. Efficiency of recovery by radio-GLC. (*a*), Sample plus internal standard; (*b*), [^{14}C]IAA.

was seen to consist of IAA using GC–MS. On the other hand, there was a smaller peak of radioactivity close to the solvent front, so merely assessing total radioactivity alone would give an erroneous impression of IAA recovery. It is necessary to show where the radioactivity is located.

Bandurski & Schulze (1974) used another method of gauging purification efficiency, in which preparative GLC was used as a final stage to purify an extract initially primed with an internal standard radioactive IAA. The peak corresponding to IAA was collected from the column effluent and a sample was then counted using liquid scintillation spectrometry. This method is adequate provided no other radioactive compounds co-chromatograph with IAA.

Radioactive internal standards are not appropriate for estimating operator variation during the mass spectrometry stages. Although we have employed internal standards of different molecules, e.g. indole acetonitrile, indole propionic acid, those are not the complete answer. In theory, IAA labelled with stable isotopes is the best type of internal standard for GC–MS analyses. The internal standard would be expected to have similar chromatographic properties to the endogenous IAA in the sample yet could be easily distinguished by inspection of the mass spectrum.

Unfortunately, problems are encountered with the use of deuterated IAA. The novel synthesis of IAA labelled by two deuterium atoms in the 2' position of the side-chain was carried out by base-catalysed exchange starting with indole acetonitrile followed by hydrolysis (Math & Hillman, unpublished). Exchange problems were noted in the glass inlet system when this deuterated IAA was used as an internal standard. (Exchange of the aryl protons by acid-catalysed deuteration was found to be impractical in that a complex mixture of indole-based polymers was produced.)

IAA labelled with ^{13}C should now be investigated as a suitable form of internal standard.

A worked example of IAA analysis using GC–MS procedures

Five hundred grams fresh weight of shoot tissue from *Phaseolus vulgaris* seedlings was extracted and extracted and purified according to the schedule in Table 1. An aliquot of [^{14}C]IAA was added to the methanolic extract as an internal standard.

The dry weight of the sample before derivatisation was 0.78 mg. This was reacted with 20 mm^3 BSA for 3 h at 60 °C and an aliquot was taken for radio-GLC. Two main peaks of radioactivity were observed, one near the solvent front and the other with an identical retention time to that of *bis*-TMSi-[^{14}C]IAA. This latter peak was used to calculate the purification efficiency, which was found to be 25%.

The remainder of the sample was subjected to GC–MS analysis on a JEOL double-focusing, single-beam MS equipped with a metal double-jet separator. The separator was heated to 150 °C. An on-line computer was available for full-scan analysis.

The MS was precalibrated with a perfluorokerosene (PFK) standard and then tuned to m/e 202. A trial injection of authentic *bis*-TMSi-IAA was carried out to establish the GC retention time. Temperature-programming from 100 to 200 °C at 15 °C min^{-1} was started 1 min after injection into a 1 m glass column of 3% OV1 on Chromosorb W, with a flow rate of 10 cm^3 min^{-1}. From the SIM trace, the retention time of the standard was found to be 8 min 12 sec.

Several injections equivalent to 10, 20, 30 and 40 ng of the *bis*-TMSi-IAA were made and the peak heights of SIM trace were used to plot a calibration curve. A definite linear response was noted. Subsequent injections of the sample revealed a peak with a similar retention time to that of the standard. The *total* sample was calculated to contain 200 ng of TMSi-IAA. As the percentage recovery of [^{14}C]IAA was 25%, then it follows that the extract contained 800 ng IAA. Co-injections of known amounts of sample and non-radioactive standard were also necessary to establish that the SIM response obeyed the calibration curve characteristics; this procedure also enables the standard and unknown to be measured under essentially identical conditions.

A full scan of the sample was also taken at the above retention time, with the scan revealing the appropriate ions corresponding to TMSi-IAA.

General precautions

(1) Avoid the use of plastic vessels, tubing and caps, otherwise prominent plasticiser peaks will appear in the mass spectrum. Line any bottle or vial caps with aluminium foil.

(2) Glassware should be thoroughly cleaned, washed with dilute hydrochloric acid, then rinsed with distilled water followed by redistilled methanol.

(3) All solvents, including those used for chromatography, should be redistilled.

(4) Thin-layer support media should be pre-washed with methanol.

(5) Regardless of what purification procedures are used, the final experimental criterion lies in the validity of the method of identification.

Problems and development

(1) It is possible that even the techniques outlined in this article may not resolve closely related structural or stereoisomers, although at present we are not aware of them.

(2) The extraction phase requires close attention and the development of suitable controls is needed. In the future, the availability of specific biosynthesis inhibitors and knowledge of the cellular location of IAA will assist in abbreviating the time taken to extract and purify the compound.

(3) The physiologist must ensure that the end result justifies the rather sophisticated and expensive means. All experiments must be carefully designed and interpreted. A considerable amount of variation could be eliminated by working with clonal plant material.

(4) Ideally, the extraction and purification stage should be simplified to one or two stages prior to critical identification and measurement. Moreover, IAA determinations will eventually have to be coupled to those of other cellular components so that *metabolic profiles* can be established for a particular treatment or condition in health or disease.

We thank Professor M. B. Wilkins, Dr B. A. Knights and Dr V. B. Math for invaluable discussions. The late Professor W. Parker and Mr D. Dance, Stirling University, are thanked for the use of the JEOL MS and computational facilities. Drs M. S. Greenwood, J. C. White, I. G. Bridges and J. Robertson all contributed through the years in developing the analysis of IAA using MS. The GC–MS facilities in the Botany Department at Glasgow were provided by grants from the SRC and J. McD. was supported by a postgraduate award from the SRC.

References
Review articles
The following five articles are recommended as excellent background reading.
Drozd, J. (1975). Chemical derivatization in gas chromatography. *Journal of Chromatography*, **113**, 303–56.
Ewing, G. W. (1975). *Instrumental Methods of Chemical Analysis*, 4th edn. Tokyo: McGraw-Hill, Kogakusha.
Gordon, A. E. & Frigerio, A. (1972). Mass fragmentography as an application of gas–liquid chromatography–mass spectrometry in biological research. *Journal of Chromatography*, **73**, 401–17.
Matucha, M. & Smolkova, E. (1976). Gas chromatography of ^{3}H- and ^{14}C-labelled compounds. *Journal of Chromatography*, **127**, 163–201.
VandenHeuvel, W. J. A. & Zacchei, A. G. (1976). Gas–liquid chromatography in drug analysis. *Advances in Chromatography*, **14**, 119–251.

Articles cited in the text
Anderson, J. A. (1968). Fractionation of indole compounds on Sephadex G-10. *Journal of Chromatography*, **33**, 536–38.

Atsumi, S., Kuraishi, S. & Hayashi, T. (1976). An improvement of auxin extraction procedure and its application to cultured plant cells. *Planta, Berlin*, **129**, 245-7.

Audus, L. J. (1972). *Plant Growth Substances*, 3rd edn, vol. 1. London: Leonard Hill.

Bandurski, R. S. & Schulze, A. (1974). Concentrations of indole-3-acetic acid and its esters in *Avena* and *Zea*. *Plant Physiology*, **54**, 257-62.

Bausher, M. G. & Cooper, W. C. (1976). HPLC separation of endogenous growth regulators in *Citrus*. In *Abstracts of 9th International Conference on Plant Growth Substances*, ed. P. E. Pilet, p. 26. Lausanne.

Bayer, M. H. (1969). Gas chromatographic analysis of acidic indole auxins in *Nicotiana*. *Plant Physiology*, **44**, 267-71.

Bentley, J. A. (1961). Extraction and purification of auxins. In *Encyclopaedia of Plant Physiology*, ed. W. Ruhland, vol. 14, pp. 501-20. Berlin: Springer.

Bertilsson, L. & Palmér, L. (1972). Indole-3-acetic acid in human cerebrospinal fluid: identification and quantification by mass fragmentography. *Science*, **177**, 74-6.

Bittner, S. & Even-Chen, Z. (1975). A GLC procedure for determining sub-nanogram levels of indol-3-yl acetic acid. *Phytochemistry*, **14**, 2455-7.

Bosin, T. R., Buckpitt, A. R. & MaicKell, R. P. (1974). Comparative gas–liquid chromatography of biologically important indoles, and their benzo(b) thiophene and 1-methylindole analogs. *Journal of Chromatography*, **94**, 316-20.

Boysen-Jensen, P. (1910). Über die Leitung des phototropischen Reizes in Avenakeimpflanzen. *Berichte der Deutschen botanischen Gesellschaft*, **28**, 118-20.

Brenner, M. L., Andersen, C. R., Ciha, A. J., Mondal, M. & Brun, W. (1976). Relationships of endogenous plant hormone content to changes in source–sink relationships. In *Abstracts of 9th International Conference on Plant Growth Substances*, ed. P. E. Pilet, pp. 49-50. Lausanne.

Bridges, I. G., Hillman, J. R. & Wilkins, M. B. (1973). Identification and localisation of auxin in primary roots of *Zea mays* by mass spectrometry. *Planta, Berlin*, **115**, 189-92.

Brook, J. L., Biggs, R. H., St John, P. A. & Anthony, D. S. (1967). Gas chromatography of several indole derivatives. *Analytical Biochemistry*, **18**, 453-8.

Burnett, D., Audus, L. J. & Zinsmeister, H. D. (1965). Growth substances in the roots of *Vicia faba* L. *Phytochemistry*, **4**, 891-904.

Darwin, C. & Darwin, F. (1880). *The Power of Movement in Plants*. London: John Murray.

Dedio, W. & Zalik, S. (1966). Gas chromatography of indole auxins. *Analytical Biochemistry*, **16**, 36-52.

De Yoe, D. R. & Zaerr, J. B. (1976). Indole-3-acetic acid in Douglas fir. Analysis by gas–liquid chromatography and mass spectrometry. *Plant Physiology*, **58**, 299-303.

Dijkman, M. J. (1934). Wuchstoff und geotropische Krümmung bei *Lupinus*. *Recueil des travaux botaniques néerlandais*, **31**, 391–450.

Dolk, H. E. (1936). Geotropie en groeistof. Dissertation, Utrecht. English translation: Geotropism and growth, F. Dolk & K. V. Thimann. *Recueil des travaux botaniques néerlandais*, **33**, 509–85.

Durley, R. C. & Kannangara, T. (1976). Use of high pressure liquid chromatography for the analysis of plant growth regulators in *Sorghum bicolor* (L) Moench. In *Abstracts of 9th International Conference on Plant Growth Substances*, ed. P. E. Pilet, pp. 81–3. Lausanne.

Elliott, M. C. & Greenwood, M. S. (1974). Indol-3-yl-acetic acid in roots of *Zea mays*. *Phytochemistry*, **13**, 239–41.

Glenn, J. L., Kuo, C. C., Durley, R. C. & Pharis, R. P. (1972). Use of insoluble polyvinylpyrrolidone for purification of plant extracts and chromatography of plant hormones. *Phytochemistry*, **11**, 345–51.

Greenwood, M. S., Shaw, S., Hillman, J. R., Ritchie, A. & Wilkins, M. B. (1972). Identification of auxin from *Zea* coleoptile tips by mass spectrometry. *Planta, Berlin*, **108**, 179–83.

Grunwald, C. & Lockard, R. G. (1970). Analysis of indole acid derivatives by gas chromatography using liquid phase OV-101. *Journal of Chromatography*, **52**, 491–3.

Grunwald, C., Vendrell, M. & Stowe, B. B. (1967). Evaluation of gas and other chromatographic separations of indolic methyl esters. *Analytical Biochemistry*, **20**, 484–94.

Hall, S. M. & Medlow, G. C. (1974). Identification of IAA in phloem and root pressure saps of *Ricinus communis* L. by mass spectrometry. *Planta, Berlin*, **119**, 257–61.

Hillman, J. R., Math, V. B. & Medlow, G. C. (1977). Apical dominance and the levels of indole acetic acid in *Phaseolus* lateral buds. *Planta, Berlin*, **134**, 191–3.

Hopping, M. E. & Bukovac, M. J. (1975). Endogenous plant growth substances in developing fruit of *Prunus cerasus* L. III. Isolation of indole-3-acetic acid from the seed. *Journal of the American Society for Horticultural Science*, **100**, 384–6.

Jamieson, W. D. & Hutzinger, O. (1970). Identification of simple naturally occurring indoles by mass spectrometry. *Phytochemistry*, **9**, 2029–36.

Kaldewey, H. & Stahl, E. (1964). Die quantitative Answertung dünn schicht-chromatographisch getrennter Auxine in *Avena* – tageslicht – test. *Planta, Berlin*, **62**, 22–38.

Kögl, F., Haagen Smit, A. J. & Erxleben, H. (1934*a*). Über ein neues Auxin ('Heteroauxin') ans Harn, XI Mitt. *Zeitschrift für physiologische Chemie*, **228**, 90–103.

Kögl, F., Haagen Smit, A. J. & Erxleben, H. (1934*b*). Über den Einfluss der Auxine auf des Wurzelwachstum und über die chemische Natur des Auxines der Graskoleoptilen. *Zeitschrift für Physiologische Chemie*, **228**, 104–12.

Larsen, P. (1955). Growth substances in higher plants. In *Modern methods of Plant Analysis*, ed. K. Paech & M. V. Tracy, vol. 3, pp. 565–625. Berlin and New York: Springer.

Mann, J. D. & Jaworski, E. G. (1970). Minimising IAA losses in plant extracts. *Planta, Berlin*, **92**, 285–91.

Miyazaki, H., Hashimoto, Y., Iwanga, M. & Kubodera, T. (1974). Analysis of biogenic amines and their metabolites by gas chromatography–chemical ionisation mass spectrometry. *Journal of Chromatography*, **22**, 575–86.

Navez, A. E. & Robinson, T. W. (1933). Geotropic curvature of *Avena* coleoptiles. *Journal of General Physiology*, **16**, 133–45.

Nitsch, J. P. (1956). Methods for the investigation of natural auxins and growth inhibitors. In *The Chemistry and Mode of Action of Plant Growth Substances*, ed. R. L. Wain & F. Wightman, pp. 3–31, London: Butterworths Scientific Publications.

Paál, A. (1919). Über phototropische Reizleitung. *Jahrbücher für wissenschaftliche Botanik*, **58**, 406–58.

Pilet, P. E. (1961). *Les Phytohormones de Croissance*. Paris: Masson.

Powell, L. E. (1964). Preparation of indole extracts from plants for gas chromatography and spectrophotofluorometry. *Plant Physiology*, **39**, 836–42.

Raj, R. K. & Hutzinger, O. (1970). Indoles and auxins. VIII. Partition chromatography of naturally occurring indoles on cellulose thin layers and Sephadex columns. *Analytical Biochemistry*, **33**, 471–4.

Reeve, D. R. & Crozier, A. (1976). Purification of plant hormones by gel permeation chromatography. *Phytochemistry*, **15**, 791–3.

Rivier, L. & Pilet, P. E. (1974). Idolyl-3-acetic acid in cap and apex of maize roots: identification and quantification by mass fragmentography. *Planta, Berlin*, **120**, 107–12.

Robertson, J., Hillman, J. R. & Berrie, A. M. M. (1976). The involvement of indole acetic acid in the thermodormancy of lettuce fruits, *Lactuca sativa* cv. Grand Rapids. *Planta, Berlin*, **131**, 309–13.

Rothert, W. (1894). Über heliotropisms. *Beiträge zur Biologie der Pflanzen*, **7**, 1–212.

Sagi, F. (1969). Silica gel or cellulose for the thin-layer chromatography of indole-3-acetic acid? *Journal of Chromatography*, **39**, 334–5.

Seeley, S. D. & Powell, L. E. (1974). Gas chromatography and detection of microquantities of gibberellins and indole acetic acid as their fluorinated derivatives. *Analytical Biochemistry*, **58**, 39–46.

Shindy, W. W. & Smith, O. E. (1975). Identification of plant hormones from cotton ovules. *Plant Physiology*, **55**, 550–4.

Söding, H. (1923). Werden von der Spitze der Haferkoleoptile Wuchshormone gebildet? *Berichte der Deutschen botanischen Gesellschaft*, **41**, 396–400.

Söding, H. (1925). Zur Kentniss der Wuchshormone in der Haferkoleoptile. *Jahrbücher für wissenschaftliche Botanik*, **64**, 587–603.

Stahl, E. & Kaldewey, H. (1961). Spuren Analyse physiologisch Aktiver, einfacher Indolderivate, Hoppe-seyl. *Zeitschrift für physiologische Chemie*, **323**, 182–91.

Steen, I. & Eliasson, L. (1969). Separation of growth regulators from *Picea abies* Karst. on Sephadex LH 20. *Journal of Chromatography*, **43**, 558–60.

Takahashi, N., Yamaguchi, I., Kono, T., Igoshi, M., Hirose, K. & Suzuki, K. (1975). Characterisation of plant growth substances in *Citrus unshiu* and their changes in fruit development. *Plant and Cell Physiology*, **16**, 1101–11.

Thimann, K. V. (1972). The natural plant hormones. In *Plant Physiology, A Treatise*, ed. F. C. Steward, vol. VIB, pp. 25–6. New York: Academic Press.

Tirimana, A. S. L. & Geevaratne, D. V. M. (1972). A sensitive and specific colour reaction for the detection of free indole in plant material. *Journal of Chromatography*, **65**, 445–6.

Went, F. W. (1928). Wuchstoff und Wachstum. *Recueil des travaux botaniques néerlandais*, **25**, 1–116.

White, J. C., Medlow, G. C., Hillman, J. R. & Wilkins, M. B. (1975). Correlative inhibition of lateral bud growth in *Phaseolus vulgaris* L. Isolation of indoleacetic acid from the inhibitory region. *Journal of Experimental Botany*, **26**, 419–24.

Whitmore, F. W. & Zahner, R. (1964). IAA synthesis by polyphenols in the extraction of *Pinus* phloem and cambial tissue. *Science*, **145**, 166–7.

D. M. A. MOUSDALE, D. N. BUTCHER & R. G. POWELL

Spectrophotofluorimetric methods of determining indole-3-acetic acid*

Introduction

An assay method which made use of the fluorescence spectrum of indole-3-acetic acid (IAA) was first suggested by Hornstein (1958). Although potentially sensitive this method lacks specificity when applied to plant extracts because (a) all compounds with an indole nucleus have fluorescent spectra similar to that of IAA, thus necessitating the separation of IAA from other acidic indoles (Dullaart, 1967) and (b) other compounds, including some phenolic acids, have fluorescent spectra overlapping that of IAA (Guilbault, 1973).

A more specific fluorimetric assay is based on the reaction of IAA with acetic anhydride, in the presence of a catalyst, to form the fluorescent

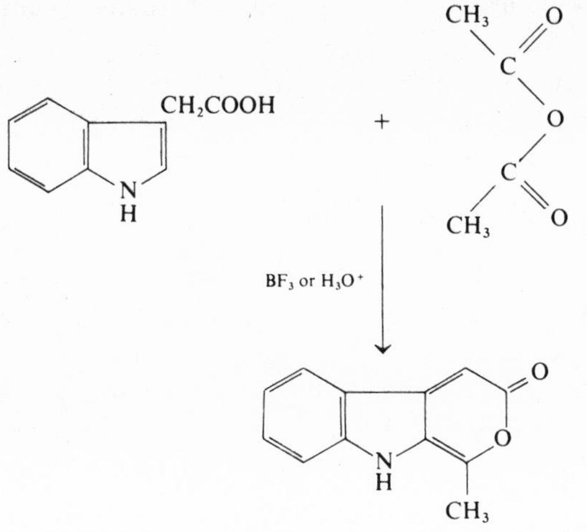

Fig. 1. The reaction of IAA with acetic anhydride to form 2-methylindole-α-pyrone.

---

* The work for this contribution was carried out at the ARC Unit for Developmental Botany, Cambridge, UK.

tricyclic derivative 2-methylindole-α-pyrone (Plieninger, Müller & Weinerth, 1964; Stoessl & Venis, 1970; Fig. 1). The reaction which is catalysed by boron trifluoride, and by strong acids, has rigorous structural requirements and few indole compounds have been shown to form fluorescent derivatives.

Experimental procedures
General

Information on the general principles of fluorescence measurements, including an appraisal of several spectrophotofluorimeters, may be found in a practical handbook written by Guilbault (1973).

It is essential that the fluorimeter used for this assay is capable of high resolution and sensitivity. In particular, it must be able to resolve the fluorescent signal of 2-methylindole-α-pyrone at 482 nm from both the Rayleigh scatter, associated with the excitation radiation around 440 nm, and the Raman scatter maximum of water at 520 nm. The instrument should be able to do this at a sensitivity setting which will produce a Raman signal of about 10–20% full-scale deflection of the recorder (Fig. 4). In addition there should be provision for recording emission and excitation spectra, for variation of the scanning speed and for accurate relation of the chart record with the wavelength. Recorders with 'strip' rather than 'X–Y' presentation of output are more convenient.

There are several suitable double monochromator recording spectrophotofluorimeters available, and the one used in these investigations was a Perkin Elmer MPF-3L.

The greatest sensitivity is achieved with the highest intensity light sources (xenon or mercury lamp). Devices which improve arc stability in the light source are desirable, though no difficulty has been experienced in this respect with the high pressure xenon lamp as stabilised in the MPF-3L instrument.

The MPF-3L instrument is fitted with the 'Ratio' detection mode as an option. The use of this mode will improve signal stability, but will inevitably reduce sensitivity.

It is clearly necessary, when attempting to achieve the greatest sensitivity, to pay careful regard to the elimination of background signals. Such unwanted responses can be reduced by ensuring that no stray light enters the cuvette chamber and that particulate matter is not present in the assay solution. Spurious fluorescence may be reduced by using chemicals of the highest available grade of purity, and in particular the methanol used should be freshly redistilled Analar grade.

All glassware should be cleaned with concentrated sulphuric acid and rinsed thoroughly in distilled water; the use of detergents is to be avoided.

Losses of IAA during the purification procedures are estimated by adding a known amount of [^{14}C]IAA, labelled in the C–1 or C–2 position, to the initial tissue extract and measuring the radioactivity in the final methanolic solution immediately prior to assay. A correction must be applied to take account of this added IAA at an appropriate stage of the calculation.

Purification procedures

Various purification procedures may be used to yield extracts suitable for assay by the pyrone fluorimetric method. With many plant tissues the following procedure, which is a modification of that devised by Knegt & Bruinsma (1973), is satisfactory:

(1) Plant tissue (1–5 g fresh weight) is frozen by immersion in liquid nitrogen, ground to a powder and extracted with aqueous methanol (80% (v/v) redistilled Analar grade methanol, 20 cm^3 g^{-1} fresh weight of tissue) at -15 °C for 90 min. A methanolic solution of [^{14}C]IAA (0.2 cm^3, approximately 4500 cpm is added at the beginning of the extraction. The suspension is filtered with suction through a Whatman No. 5 filter paper and the residue washed with methanol (10 cm^3). The extraction of more than 20 g of tissue is often unsuccessful because of the resulting higher concentrations of interfering substances.

(2) This initial methanolic extract is evaporated to an aqueous residue under reduced pressure and the residue is dissolved in 15 cm^3, 0.5 mol dm^{-3} K$_2$HPO$_4$ (pH 8.5) buffer. Filtration may be necessary at this stage.

(3) The buffer is shaken with 2×20 cm^3 freshly redistilled diethyl ether and the aqueous phase is separated and acidified to pH 3 with 3 mol dm^{-3} H$_3$PO$_4$.

(4) The aqueous phase is then shaken with another 20 cm^3 diethyl ether. The ether phase is separated and shaken with 10 cm^3 0.05 mol dm^{-3} K$_2$HPO$_4$ buffer (pH 8.5). The resulting aqueous phase is acidified to pH 3.0 with 0.3 mol dm^{-3} H$_3$PO$_4$ and again shaken with 20 cm^3 ether.

(5) The final ether fraction is then separated and evaporated at room temperature with a stream of nitrogen. The dry residue is dissolved in 2 cm^3 redistilled methanol. An aliquot (0.2 cm^3) of this solution is taken for radioactive assay to give the ^{14}C recovery during the purification procedure and the remainder is used in the fluorimetric assay. The recovery efficiency is usually between 60 and 75%.

Removal of contaminants which cause light scattering and reaction inhibition. In some cases the above procedure yields extracts containing contaminating substances which may seriously interfere with the assay. Two types of contaminant have been distinguished: those causing (*a*) light scattering and (*b*) reaction inhibition.

(i) *Light scattering.* In some instances (e.g. with stem tissue of *Helianthus annuus*) the addition of water to stop the reaction gives rise to an opalescence in the reaction mixture. This causes a large amount of light scattering from the excitation beam, which often obscures the signal caused by the 2-methylindole-α-pyrone. Unlike the decay of the 'pyrone' fluorescence (see below) the decay of the signal resulting from light scattering is inconsistent and unpredictable. Although the precise nature of the contaminants causing opalescence is unknown the difficulty can be avoided by the addition of an extra step in the purification procedure. The aqueous phase in 0.05 mol dm^{-3} phosphate buffer, from the modified Knegt and Bruinsma procedure, is applied to a 1×5 cm column of insoluble polyvinylpyrrolidone (PVP; GAF Ltd, Calder Street, Manchester; approximately 80–120 mesh) and eluted with the same buffer. The fraction containing IAA (5–25 cm^3) is acidified with 0.3 mol dm^{-3} H$_3$PO$_4$ (pH 3.0) and extracted with 25 cm^3 redistilled ether. The recovery of IAA from the column is greater than 99%.

Eliasson, Strömquist & Tillberg (1976) have reported that light scattering resulting from contaminants in extracts from *Betula verrucosa* may be avoided by the addition of ethanol, after stopping the assay reaction with water. However, this often results in the introduction of an undesirably high level of background fluorescence.

(ii) *Reaction inhibition.* The phenomenon of reaction inhibition, which was first described by Stoessl & Venis (1970), manifests itself as a reduction in the amount of fluorescence per unit of IAA present in the extract. Although a certain amount of the inhibition can be corrected for (see below), in severe cases special purification procedures have to be adopted, e.g. with extracts from etiolated tissues of *Zea mays* and *Triticum vulgare*. Attempts to remove the inhibiting components from these extracts by column chromatography on Sephadex G-10 or PVP, or by thin-layer chromatography, have been unsuccessful. Furthermore, the use of electrophoresis for the removal of inhibitors from extracts of a number of other species has been only partially successful (Eliasson *et al.*, 1976).

Extracts from *Z. mays* and *T. vulgare* suitable for assay have been prepared using a procedure modified from that of Atsumi, Kuraishi & Hayashi (1976). This procedure is as follows:

(1) The initial methanolic extract from 1–5 g fresh weight of tissue is evaporated to an aqueous residue, dissolved in 50 cm³ of 80% saturated ammonium sulphate solution and adjusted to pH 3.5 with HCl (20% (v/v)).

(2) The solution is filtered and shaken with 2×60 cm³ redistilled dichloromethane.

(3) The combined dichloromethane fractions are then re-extracted by shaking with 20 cm³ 0.05 mol dm⁻³ K_2HPO_4 buffer (pH 8.5).

(4) The aqueous phase is separated and immediately acidified to pH 3.0 with 0.3 mol dm⁻³ H_3PO_4 and then shaken with 2×30 cm³ dichloromethane.

(5) The combined dichloromethane fractions are evaporated to dryness under reduced pressure at room temperature, and the residue is dissolved in 2 cm³ redistilled methanol and assayed. The recovery efficiency is monitored as before.

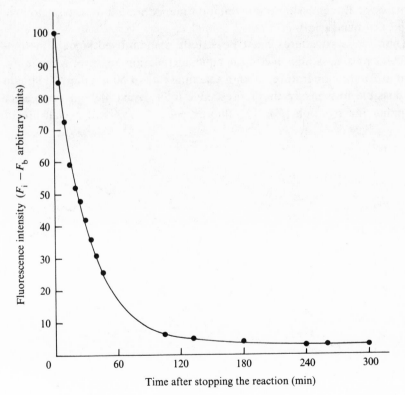

Fig. 2. Decay of 2-methylindole-α-pyrone fluorescence. 20 ng IAA were reacted with acetic anhydride, with trifluoroacetic acid as catalyst. After stopping the reaction, the reaction mixtures were left at 23 °C in the light. F_i = 'initial' fluorescence; F_b = background fluorescence.

Assay procedure

(1) The methanolic solution (0.2 cm³), containing IAA, is placed in a clean dry test tube (120 mm×12 mm) and the methanol is carefully evaporated in a stream of nitrogen at 40 °C.

(2) After cooling the reaction tube in an ice:water mixture, the reaction is initiated by adding 0.1 cm³ acetic anhydride followed by 0.1 cm³ of the catalyst (trifluoroacetic acid or 60% (w/v) perchloric acid).

(3) After the appropriate reaction time (the optimum reaction times are 10 min for perchloric acid or 15 min for trifluoroacetic acid), the reaction is stopped by the addition of 4 cm³ distilled water. The aqueous solution is mixed and transferred immediately to a silica fluorescence cuvette. With the excitation wavelength fixed at 440 nm, the emission spectrum is recorded between 450 and 540 nm. The excitation wavelength of 440 nm is chosen in preference to 450 nm (an excitation maximum) in order to reduce the Rayleigh scatter signal in the region of 482 nm. The slit width controls of the emission and excitation monochromators are set to give a spectral band width of 5 nm.

The assay procedures must be strictly standardised since 2-methyl-indole-α-pyrone is unstable in the aqueous reaction mixture, when in the light at room temperature. With a scanning rate of 50 nm min⁻¹ it should be possible to measure the fluorescence intensity at 480 nm 1 min after stopping the reaction (F_i). The fluorescence decays with a half-life of

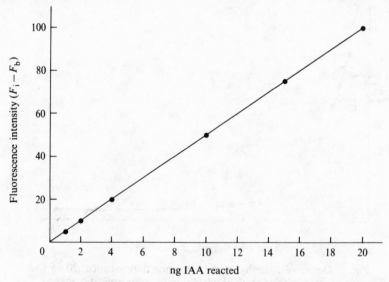

Fig. 3. Calibration graph for the pyrone fluorimetric assay (perchloric acid as catalyst; reaction time, 10 min). F_i = 'initial' fluorescence; F_b = background fluorescence.

approximately 15 min and reaches a stable background level after 4 h (Fig. 2). By noting this background level (F_b) and subtracting it from the 'initial' fluorescence (F_i), a measure of the fluorescence intensity resulting from the 2-methylindole-α-pyrone formed in the reaction can be obtained. This procedure is more satisfactory than the use of a blank (prepared by adding water before acetic anhydride and catalyst) because it gives a more accurate estimation of the stable fluorescence not attributable to 2-methylindole-α-pyrone. Using these procedures with standard methanolic solutions of authentic IAA, there is a straight line relationship between the fluorescence intensity caused by 2-methylindole-α-pyrone ($F_i - F_b$) and the amount of IAA reacted (Fig. 3).

In general, perchloric acid is a superior catalyst to trifluoroacetic acid for measuring small amounts of IAA (less than 10 ng), since the emission spectra can be resolved at lower levels of IAA (Fig. 4). With larger amounts of IAA, trifluoroacetic acid is also satisfactory.

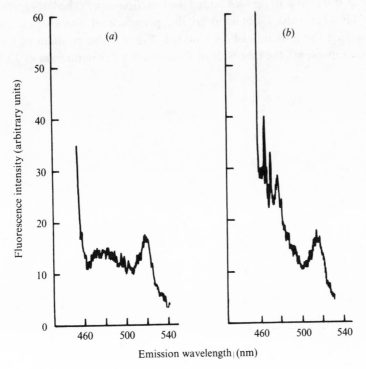

Fig. 4. The fluorescence emission spectrum of the reaction product from 1 ng IAA using (a) perchloric acid, (b) trifluoroacetic acid as catalyst. Excitation wavelength: 440 nm. The peak at 520 nm is a Raman scatter peak.

The scanning procedure outlined here is preferable to the measurement of fluorescence at a single wavelength used by previous workers, since it permits the detection of light scattering and may confirm that the emission maximum, if resolvable, corresponds to that of IAA.

The estimation of IAA in plant extracts by the fluorimetric assay is frequently complicated by the presence of contaminants which inhibit the reaction between IAA and acetic anhydride (Stoessl & Venis, 1970). As stated previously, it causes a reduction in the amount of fluorescence per unit of IAA in the presence of a plant extract. This is not caused by fluorescence quenching, since the effect cannot be reproduced by adding the plant extract to a completed reaction mixture. However, as long as the inhibition is not too severe, a correction can be made using the procedure suggested by Stoessl & Venis (1970). The apparent amount of IAA in an aliquot of a final methanolic extract is calculated by reference to a calibration graph (Fig. 3). Assays are then performed with appropriate amounts of IAA standard solutions (in duplicate tubes) mixed with the extract (e.g. 0, 2, 5 and 10 ng IAA added per reaction tube). The fluorescent intensity $(F_i - F_b)$ values obtained in the presence of the extract are plotted against the amounts of IAA added (Fig. 5). The gradient of this graph then represents the intensity of fluorescence per nanogram of IAA

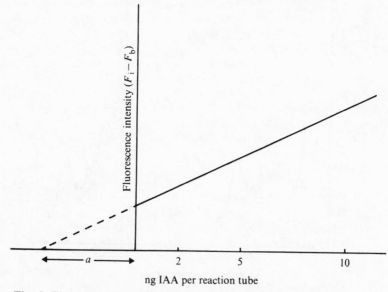

Fig. 5. The relationship between fluorescence intensity and IAA concentration in the presence of a plant extract. Distance a represents endogenous IAA. F_i = 'initial' fluorescence; F_b = background fluorescence.

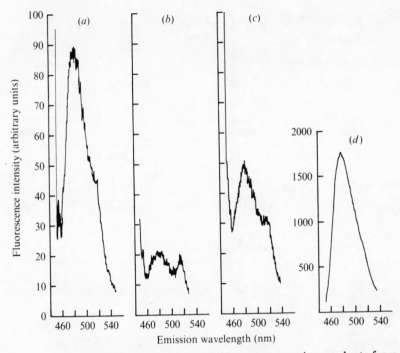

Fig. 6. The emission spectra of the pyrone reaction products from (*a*) 10 ng IAA; and extracts of (*b*) etiolated *Avena sativa* coleoptile, (*c*) etiolated *Pisum sativum* stem and (*d*) *Zea mays* seed.

present. In Fig. 5 it can be seen that the distance *a* between the origin and the intercept on the abscissa represents the amount of endogenous IAA. Alternatively the endogenous IAA, *a*, may be calculated from the relationship

$$a = \frac{(F_i - F_b)\ \text{sample}}{\text{gradient}}$$

Clearly the correction which is made by either procedure is dependent on the gradient of the line in Fig. 5. The reliability of the correction diminishes when the gradient is low.

The emission spectra of the pyrone reaction products from 10 ng IAA (*a*) and sample extracts (*b*, *c* and *d*) are shown in Fig. 6. Estimations of the free-IAA levels in a range of tissues, obtained using the assay procedure, are given in Table 1.

Table 1. *Free IAA levels measured in plant tissues by the pyrone fluorimetric assay*

Plant species	Tissue	Age	Free IAA content(ng g^{-1} fresh weight) Duplicate extractions
Lycopersicum esculentum	Stem	5 weeks[a]	14.8 15.4
Helianthus annuus	Stem	5 weeks[a]	30.0 40.0
Sinapis alba	Cotyledon	4 weeks[a]	8.1 8.3
Pisum sativum	Etiolated stem (third internode)	6 days[b]	34.0 35.2
Avena sativa	Coleoptile (etiolated)	3½ days[b]	20.0 22.2
	Mesocotyl (etiolated)	3½ days[b]	9.8 12.8
Nicotiana rustica	Stem	7½ weeks[a]	19.6 20.4
Zea mays	Seed (unimbibed)		540 600

[a] Plants grown in a heated glasshouse.
[b] Plants grown in a dark room at 25 °C.

Assay specificity

As stated earlier, a few indole compounds other than IAA react with acetic anhydride to form fluorescent derivatives similar to 2-methylindole-α-pyrone (Table 2). The fluorescence excitation and emission maxima of the reaction products from IAA and five other indole compounds are shown in Table 2. An α-pyrone from N-methylindole-3-acetic acid has also been described (Dorofeenko & Korobkova, 1968). However, these are unlikely to present problems in the assay for the following reasons.

(1) An examination of the fluorescence spectra provides a method of distinguishing between 2-methylindole-α-pyrone and the other fluorescent derivatives. Further evidence regarding the identity of the fluorescent derivatives may be obtained if excitation maxima are recorded with the emission wavelength fixed at 480 nm (Table 2). For this, relatively large amounts of IAA are required. The maxima given in the table are uncorrected and likely to differ according to the instrument being used.

Table 2. *The emission and excitation maxima of the fluorescent derivatives formed from indoles in the pyrone reaction*

1 μg of each compound was reacted with acetic anhydride with trifluoroacetic acid as catalyst. The spectra were recorded in aqueous solution using a Perkin–Elmer MPF-3L fluorescence spectrophotometer fitted with a xenon light source.

Indole reactant	Excitation maxima (nm)	Emission maximum (nm)
Indole-3-acetic acid (IAA)	275, 310, 450	482
5-Hydroxyindole-3-acetic acid	275, 308, 466	487
5-Methoxyindole-3-acetic acid	290, 330, 470	522
Indole-3-acetonitrile	280, 332, 450	505
Indole-3-acetamide	280, 317, 470	503
Indole-3-acetone	280, 327, 423	495

(2) The emission maxima of the fluorescent derivatives of indole-3-actonitrile, indole-3-acetamide, indole-3-acetone and 5-methoxyindole-3-acetic acid are sufficiently different from 480 nm (the emission maximum of 2-methylindole-α-pyrone) to make their fluorescence at this wavelength small (Table 3).

(3) Indole-3-acetonitrile, indole-3-acetamide and indole-3-acetone are neutral compounds and should be removed by the extraction procedure.

(4) Many derivatives of IAA with ring substituents are likely to give rise to fluorescent compounds with acetic anhydride, but they have rarely been found in plants. The one exception, to date, is 4-chloroindole-3-acetic acid which has been reported in immature pea seeds (Marumo, Hattori, Abe & Munakata, 1968).

Table 3. *The relative fluorescence intensities of the known fluorescent derivatives formed in the reaction conditions of the pyrone assay*

The emission wavelength was set at 480 nm and the excitation wavelength at 440 nm. 1 μg of each compound was reacted for 15 min at 0 °C. with acetic anhydride using trifluoroacetic acid as a catalyst.

Indole reactant	Relative fluorescence intensity
IAA	100
5-Hydroxyindole-3-acetic acid	54.3
5-Methoxyindole-3-acetic acid	0.2
Indole-3-acetonitrile	1.9
Indole-3-acetamide	2.0
Indole-3-acetone	0.6

The high specificity of the assay is indicated by the fact that the following indoles do not form detectable amounts of fluorescent derivatives in the α-pyrone assay: indole-3-aldehyde, indole-3-lactic acid, indole-3-propionic acid, indole-3-butyric acid, indole-3-glyoxylic acid, indole-3-glycollic acid, indole-3-acetylaspartic acid, indole-3-propionamide, indole, skatole, tryptamine, tryptophan, indoxylacetate (Stoessl & Venis, 1970), indole-3-acrylic acid, indole-3-pyruvic acid, indole-3-acetaldehyde (as the bisulphite addition compound), indole-3-ethanol, indole-3-glyoxamide, the ethyl ester of IAA, indole-2-carboxylic acid and its ethyl ester, indole-5-carboxylic acid and the ethyl ester of indole-1-carboxylic acid.

In addition derivatives with fluorescent spectra similar to 2-methylindole-α-pyrone are not formed from the following compounds: anthranilic acid, nicotinic acid, picolinic acid, quinaldic acid, caffeic acid, ferulic acid, sinapic acid, 2,4-dichlorophenoxyacetic acid, naphthyl-1-acetic acid, naphthoxy-2-acetic acid, phenylacetic acid, p-hydroxyphenylacetic acid, gibberellic acid, abscisic acid and N^6-iso-pentenyladenine.

Appraisal of the method

The particular advantages of the technique in the measurement of endogenous IAA levels in plant tissues are:

(1) The extraction, purification and determination may be completed within 8 h.

(2) The sensitivity is high and $1 \, ng \, g^{-1}$ of IAA can be detected in favourable tissues.

(3) It may be used for a wide variety of tissues if adequate purification procedures are selected and care is taken to confirm that the fluorescence being measured results only from 2-methylindole-α-pyrone. This latter point can only be ensured by using a scanning procedure.

References

Atsumi, S., Kuraishi, S. & Hayashi, T. (1976). An improvement of auxin extraction procedure and its application to cultured plant cells. *Planta, Berlin*, **129**, 245–8.

Dorofeenko, G. N. & Korobkova, V. G. (1968). Synthesis of substituted indole(2,3-c)pyrylium salts, indole-α-pyrones and indole-α-pyridones. *Chemistry and Industry*, 1848.

Dullaart, J. (1967). Quantitative estimation of indoleacetic acid and indole carboxylic acid in root nodules and roots of *Lupinus luteus* L. *Acta botanica neerlandica*, **16**, 222–30.

Eliasson, L., Strömquist, L.-H. & Tillberg, E. (1976). Reliability of the indole-α-pyrone fluorescence method for indole-3-acetic acid determination in crude plant extracts. *Physiologia Plantarum*, **36**, 16–19.

Guilbault, G. G. (1973). *Practical fluorescence. Theory, methods and techniques.* New York: Marcel Dekker.

Hornstein, I. (1958). Spectrophotofluorometry for pesticide determinations. *Journal of Agricultural and Food Chemistry*, **6**, 32–4.

Knegt, E. & Bruinsma, J. (1973). A rapid, sensitive and accurate determination of indolyl-3-acetic acid. *Phytochemistry*, **12**, 753–6.

Marumo, S., Hattori, H., Abe, H. & Munakata, K. (1968). Isolation of 4-chloroindolyl-3-acetic acid from immature seeds of *Pisum sativum*. *Nature, London*, **219**, 959–60.

Plieninger, H., Müller, W. & Weinerth, K. (1964). Indole-α-pyrone and indolo-α-pyridone. *Chemische Berichte* **97**, 667–81.

Stoessl, A. & Venis, M. A. (1970). Determination of submicrogram levels of indole-3-acetic acid. A new, highly specific method. *Analytical Biochemistry*, **34**, 344–51.

D. R. REEVE & A. CROZIER

The analysis of gibberellins by high performance liquid chromatography

Summary

High performance liquid chromatography is a powerful separatory technique, well suited to the analysis of gibberellins in plant extracts. However, as its sample capacity is limited the potential of high performance liquid chromatography is best exploited by restricting its application to extracts of relatively high purity. Consequently the techniques employed for the purification of gibberellins prior to high performance liquid chromatography are as important as the high performance liquid chromatography procedures themselves. The rationale behind the way in which high performance liquid chromatography is used, and its relationship with other purification procedures and mass spectrometry is discussed in detail and points illustrated by reference to the analysis of [^{3}H]gibberellin metabolites in extracts of *Phaseolus coccineus* seedlings. Possible uses of high performance liquid chromatography in the analysis of endogenous gibberellins are also outlined.

Introduction

Primarily as a result of the studies of MacMillan and co-workers combined gas chromatography–mass spectrometry (GC–MS) has become a procedural cornerstone in the rigorous analysis of gibberellins (GAs) in plant extracts (Binks, MacMillan & Pryce, 1969). The effectiveness of this technique is, however, dependent upon the concentration of GA in the extract under investigation as 5% purity is usually necessary if a positive characterisation is to be achieved (MacMillan, 1972), although in favourable circumstances and with the availability of sophisticated signal processing facilities much lower concentrations can be analysed (see Gaskin & MacMillan, this volume). This requirement for purity is easily met in extracts of the fungus, *Gibberella fujikuroi*, which is a prolific producer of GAs. However, with the possible exception of immature seed, the concentration of GAs in most higher plant tissues is several orders of magnitude lower than in *Gibberella*, at approximately 1 part in 10^9–10^{10}. Consequently higher plant extracts typically consist of small amounts of GAs contaminated with vast quantities of impurities. Characterisation of the GAs in such extracts presents formidable technical problems because, even after employing a series of purification steps, the minimum level of purity required for GC–MS analysis is only rarely achieved. New avenues must therefore be explored if GAs from higher plants are to be identified on a more routine basis, and in this respect recent advances in chromatographic technology and the development of high performance liquid chromatography (HPLC) have much to offer.

HPLC has been applied to a diversity of problems, ranging from the analysis of opium alkaloids and hashish extracts to the assay of insecticides and herbicides (Done, Knox & Loheac, 1974). High column efficiencies, rapid speeds of analysis, the availability of numerous supports offering markedly different selectivities, operation at ambient temperature, and ease of sample recovery all contribute to the overall effectiveness of the technique (Done, Kennedy & Knox, 1972; Knox, 1974). However, it should be noted that in order to achieve high column efficiencies most commercial chromatographs are designed around so-called 'analytical' columns with a 2–5 mm bore. The sample capacity of these columns rarely exceeds 500 μg and they are therefore of very limited value in the analysis of GAs in impure extracts. The considerable potential of HPLC can be exploited only when it is applied to extracts of relatively high purity. This is a crucial point and presupposes the availability of ancillary procedures capable of increasing the purity of the GAs to a level compatible with the restricted sample capacity of HPLC. To overcome this restraint we have developed a 'preparative' high performance liquid chromatograph that offers 40 times the sample capacity of commercial 'analytical' systems without unduly compromising efficiency and speed of analysis (Reeve, Yokota, Nash & Crozier, 1976). Because of its high sample capacity the 'preparative' chromatograph can easily accommodate relatively crude plant extracts and improve purity to a degree where 'analytical' HPLC and subsequent mass spectrometry become feasible (Crozier & Reeve, 1977).

These techniques have been used in our studies of GA metabolism in *Phaseolus coccineus* seedlings in which it is necessary to characterise [³H]GA metabolites routinely. The tissues are extracted with methanol and partially purified by partitioning procedures, gel permeation and charcoal adsorption chromatography. The semi-purified extracts then undergo 'preparative' HPLC followed by 'analytical' HPLC. At this stage the degree of purity is well in excess of 5% and satisfactory mass spectra of the GA metabolites can be obtained. This report will describe these techniques in detail and outline the quantitative contribution of each step to the overall efficiency of the isolation procedure. In addition the possible application of HPLC to the quantitative and qualitative analysis of endogenous GAs will be discussed.

Extraction and partitioning procedures

The procedures used are outlined in Fig. 1. Tissue is macerated and extracted three times with an excess of cold methanol. Studies with a range of [³H]GAs indicate that the methanol removes all the precursor and its metabolites from the plant tissue. The combined methanolic extracts are reduced to the aqueous phase *in vacuo* and the aqueous residue diluted at least two-fold with pH 8.0, 0.5 mol dm^{-3} phosphate buffer. This stabilises the pH and guarantees a minimum ionic strength during the ensuing partitioning procedures. At pH 8.0 the aqueous phase is sufficiently basic to retain even the less polar GAs when partitioning against light petroleum (b. pt 60–80 °C) yet not so basic as to risk GA isomerisation. More impurities would be removed from the aqueous phase if it were partitioned against a stronger solvent, such as ethyl acetate or diethyl ether, but this would also result in the loss of significant quantities of non-polar GAs.

After partitioning against light petroleum the buffer phase can be further purified by slurrying with insoluble polyvinylpyrrolidone (PVP) (Glenn, Kuo, Durley & Pharis, 1972) before acidification to pH 2.5 and extraction with $5 \times \frac{2}{5}$ volumes of ethyl acetate. At this pH, the partition coefficients are such that the bulk of the free GAs will be removed by the ethyl acetate (Durley & Pharis, 1972). The tetrahydroxy compound, GA$_{32}$, is the only known free GA that will be retained by the buffer to any extent (Yamaguchi *et al.*, 1970). Our experiments with [³H]GAs indicate that certain GA conjugates also migrate into the ethyl acetate. It is difficult to assess what proportion of the conjugated GA fraction this represents as little is known at present about their partitioning behaviour. However, it is possible to extract conjugates from the acidified aqueous phase with *n*-butanol.

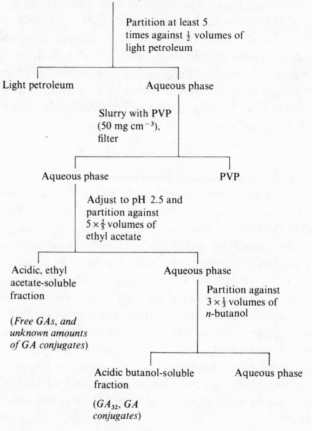

Macerate tissue and extract 3 times with an
excess of cold methanol. Combine methanolic
extracts and reduce to the aqueous phase *in
vacuo*. Add at least an equivalent volume of
pH 8.0, 0.5 mol dm^{-3} phosphate buffer and
if necessary adjust extract to pH 8.0

Partition at least 5
times against $\frac{1}{2}$ volumes of
light petroleum

Light petroleum   Aqueous phase

Slurry with PVP
(50 mg cm^{-3}),
filter

Aqueous phase   PVP

Adjust to pH 2.5 and
partition against
$5 \times \frac{2}{3}$ volumes of
ethyl acetate

Acidic, ethyl   Aqueous phase
acetate-soluble
fraction   Partition against
                                    $3 \times \frac{1}{3}$ volumes of
(Free GAs, and   *n*-butanol
unknown amounts
of GA conjugates)

Acidic butanol-soluble   Aqueous phase
fraction

(GA$_{32}$, GA
conjugates)

Fig. 1. Flow diagram of extraction and partitioning procedures.

Reducing the acidic, ethyl acetate-soluble fraction to dryness will subject the extract to undesirably severe hydrolysis conditions unless precautions are taken to prevent the formation of an aqueous residue. This
can be achieved by either (*a*) drying the wet solvent with anhydrous sodium
sulphate to reduce the water content to below 8%, which corresponds to
the composition of the ethyl acetate/water azeotrope, or (*b*) evaporating
the solvent *in vacuo* to approximately a quarter of its original volume
before adding a large excess of toluene and removing the remaining
aqueous ethyl acetate as the toluene/ethyl acetate/water ternary azeo-

trope. In practice it is difficult to dry ethyl acetate thoroughly with anhydrous sodium sulphate so both procedures should be performed. Large volumes are best dealt with by cooling to below $-10\,°C$, filtering off the frozen aqueous residue and removing any remaining moisture with a small quantity of anhydrous sodium sulphate.

During partitioning, intractable emulsions often form. Four simple procedures can be employed to overcome this problem:

(i) Dilute the emulsifying agent by adding more buffer.

(ii) If the emulsifying agent is proteinaceous it can be precipitated by adding large amounts of sodium chloride. This procedure will, however, tend to increase the amount of material that migrates into the ethyl acetate.

(iii) A particularly effective step is to remove the emulsifying agents by filtration through cellulose powder (Solka floc).

(iv) If all else fails, break the emulsion by centrifugation.

Group separatory procedures

Because of the high dry weights encountered, the initial purification of the acidic, ethyl acetate-soluble extract requires a chromatographic step with a high sample capacity, and, although the effectiveness will be limited by the wide structural diversity of the individual GAs, it should also have the ability to separate the GAs as a group from other components in the extract. In practice the greatest reductions in dry weight are obtained when more than one group separatory procedure is employed and when the individual techniques display distinctly different separatory mechanisms. For example, as a first step we use gel permeation chromatography (GPC) where the separations are based on molecular size differences (Reeve & Crozier, 1976). This is followed by exploitation of the extraordinary reverse phase effects of charcoal adsorption chromatography. These two techniques bring about a considerable increase in sample purity without being unduly time consuming.

GPC utilises two 2.5×100 cm glass columns connected in series, packed with Bio-Beads SX-4 (Bio-Rad Laboratories Ltd, 27 Homesdale Road, Bromley, Kent) and eluted with tetrahydrofuran (THF) delivered by a micrometering pump at a rate of $2.0\ cm^3\ min^{-1}$. This is the maximum flow rate the porous polystyrene beads can tolerate without excessive compression of the bed. Material eluting from the column is monitored by means of a Frensel type differential refractometer linked to a chart recorder. The gel has an operating range of 0–1500 molecular weight units and solutes elute in order of decreasing molecular size. The sample capacity is high and can be readily realised because of the excellent solubilising power of THF, 1.5 cm³ of which will dissolve up to 1.0 g of an acidic, ethyl acetate extract. The recoveries estimated with a range of [³H]GAs are in excess of 90%. The absence of adsorption effects ensures that even with the most impure extracts all the components will be eluted by the passage of a solvent volume corresponding to the total volume of the column, and this allows the system to be used repeatedly without fear of sample overlap.

Fig. 2 shows the data obtained after GPC of the acidic, ethyl acetate-soluble fraction obtained from *Phaseolus coccineus* seedlings extracted 8 h after the application of [³H]GA₁₄. There are two distinct zones of radioactivity. The peak eluting first corresponds to the GA conjugate fraction that partitions into ethyl acetate at pH 2.5. The more retained zone of radioactivity represents free GAs. The refractive index monitor trace gives an estimate of the distribution of dry weight with respect to molecular size, and by collecting the free GA zone a two- to three-fold increase in sample purity will be obtained.

Charcoal adsorption chromatography of the semi-purified free GA zone from GPC is carried out on a 20×120 mm column of charcoal–celite (1:2). The sample is dissolved in 1–2 cm³ of 20% (v/v) aqueous acetone and

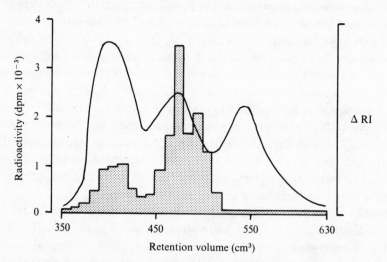

Fig. 2. GPC of the acidic, ethyl acetate-soluble fraction from 200 light-grown *Phaseolus coccineus* seedlings 8 h after the application of 15 μCi of [³H]GA₁₄. The refractive index (RI) trace was obtained with an on-stream differential refractometer linked to a chart recorder. Successive 20 cm³ fractions were collected and 1/1000 aliquots of each used to estimate radioactivity (shaded histogram) by liquid scintillation counting.

applied to the top of the column. Weakly adsorbed solutes are eluted with 100 ml of 20% aqueous acetone which is equivalent to four column volumes. The GAs are then removed with 200 cm^3 of acetone giving a 75–85% recovery. The adsorption capacity of charcoal is high and a column of the dimensions described can easily accommodate up to 500 mg of extract. Charcoal chromatography results in an approximately four-fold reduction in sample weight with *Phaseolus* seedling extracts, and should follow GPC as it removes a THF polymerisation product which forms within hours of the distillation of the solvent from its stabiliser.

Other preparative group separatory procedures have been used for the purification of GAs. Although PVP adsorption chromatography can significantly reduce the dry weight of an extract (Glenn *et al.*, 1972), very low column efficiencies and long analysis periods are associated with this technique. Further it involves the use of aqueous solvents which are undesirable both from the point of view of sample recovery and because of the risk of GA rearrangements (Pryce, 1973). It is therefore safer, easier and almost as effective to make use of the PVP slurry treatment described in the previous section. A Sephadex G-10 column eluted with 0.1 mol dm^{-3} pH 8.0 phosphate buffer will retain GAs by virtue of uncharacterised adsorption phenomena and can be of value as a purification tool (Crozier, Aoki & Pharis, 1969). However, as this procedure also involves exposure of the extract to mildly alkaline conditions for several hours, with attendant risks of degradation, it should also be avoided if the required degree of purity can be achieved without recourse to a hydrolytic environment.

'Preparative' high performance liquid chromatography

After purification of *Phaseolus* seedling extracts by GPC and charcoal chromatography, the concentration of individual radioactive GA metabolites rarely exceeds 0.1%. Even if additional group separatory procedures were to be employed the purity would not approach the level required for GC–MS. This can be achieved only through the use of analytical techniques which separate the individual GAs from one another. If large numbers of samples are to be routinely processed it is imperative that the first analytical step should be both efficient and rapid. It must also provide as much information as possible on the nature of the metabolites, otherwise an unwieldy number of subfractions will be generated by the ensuing procedures. The additional requirements for high sample capacity and good recoveries restrict the choice of technique to liquid chromatography systems based on wide diameter columns and high stationary phase loadings. In such systems the sample capacity is usually gained at the expense of efficiency and speed of analysis, which, although of secondary importance, ought not to be entirely sacrificed. A further requirement is that the system should have a high peak capacity to enable it to handle the diversity of compounds present in plant extracts. This means that the liquid chromatography procedure chosen must be compatible with solvent programming techniques.

We have developed a 'preparative' high performance liquid chromatograph which meets these criteria (Reeve et al., 1976). The instrument is fitted with an on-stream, homogeneous radioactivity monitor to facilitate the detection and analysis of [³H]GAs (Fig. 3). The chromatograph uses a 10×450 mm analytical column packed with either Partisil 10 or Partisil 20 silica gel particles (Whatman Ltd, Springfield Mill, Maidstone, Kent) on which are loaded a 40% (v/w) 0.5 mol dm⁻³ formic acid stationary phase. A dual pump gradient generator delivers mixtures of hexane and ethyl acetate saturated with 0.5 mol dm⁻³ formic acid to the analytical column via a pulse dampening network, a stationary phase trap, and a precolumn. Samples are dissolved in the mobile phase and introduced into the analytical column via a six port sample valve. Solvent emerging from the column is directed to the flow cell of a UV monitor before entering a stream splitter which subtracts a preset portion of the column eluant and restores the original flow rate with a 'make-up' solvent of ethyl acetate:toluene (1:1::v:v). After the addition of scintillation cocktail supplied from a metering pump the mixture is cooled to −5 °C and passed through a spiral

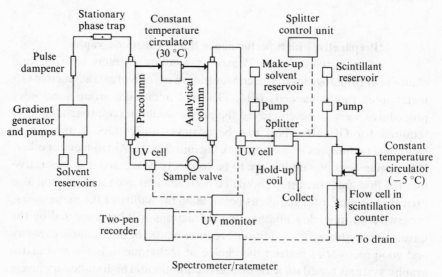

Fig. 3. The 'preparative' HPLC with UV and radioactivity monitors.

glass flow cell positioned between the photomultiplier tubes of a manual scintillation counter (Laboratory Impex, PO Box 14, Twickenham, Middlesex). The output is processed by a spectrometer/ratemeter (Laboratory Impex, PO Box 14, Twickenham, Middlesex) and displayed along with the UV-absorbance trace on a dual channel recorder.

A 10×450 mm column packed with Partisil 10 and eluted at a flow rate of $5.0 \, cm^3 \, min^{-1}$, which corresponds to a solvent linear velocity of 1.5 mm s^{-1}, generates up to 3800 theoretical plates for a solute with a capacity factor (k') of 1.2. Thus the plate height and the speed of analysis can be calculated as 0.12 mm and 1.1 effective plates per second, respectively. Depending upon solvent composition a column inlet pressure of 140–200 psi is required. Individual solutes of up to 10 mg in weight and 330 mm^3 in volume can be chromatographed without significant loss in resolution, while multicomponent samples weighing up to 100 mg can be analysed without overloading the system. Recovery from the column is better than 95% for the wide range of compounds tested. These performance figures represent a considerable improvement in both efficiency and speed of analysis, when compared with classical liquid chromatography techniques used for the separation of GAs. The system is some ten times faster and twenty times more efficient than the silica gel partition column of Powell & Tautvydas (1967) from which it was derived.

The factors which restrict the performance of preparative columns of this type can be summarised as follows. Because of the exceptionally high stationary phase content, limiting stationary phase mass transfer contributes significantly to the plate height and is most acute for solutes with k' values in the range 1.5 to 2.5. By using supports of small particle diameter it is possible to reduce the diffusional path-lengths and bring about a spectacular increase in both the efficiency and the speed of analysis of the column. Provided supports with a narrow particle size range and uniform pore geometry are available, particles as small as 10 μm can be used without recourse to excessively high column inlet pressures. The instability of the stationary phase on the support, which is a consequence of the miscibility of ethyl acetate and formic acid, is another major cause of bandspreading. Precautions that can be taken to reduce this problem include the use of an effective stationary phase trap in the solvent delivery line and a precolumn, which, along with the analytical column, is maintained at $30 \pm 0.05 \, °C$ to ensure equilibration of the incoming mobile phase with the stationary phase. When not in use the columns should be stored at constant temperature with the mobile phase displaced by nitrogen, in order to encourage the stationary phase to distribute itself uniformly throughout the column by migration in the vapour state.

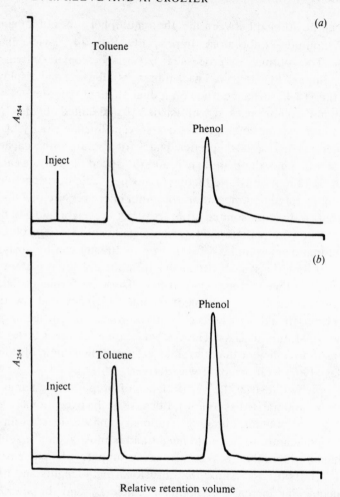

Fig. 4. The influence of sample introduction on peak symmetry. Solutes introduced off-column with a 330 mm³ six-port valve via (a) a standard column inlet fitting and (b) a modified inlet fitting to confine the sample to the axial centre of the column – see Fig. 5. Column: 10×450 mm Partisil 20. Stationary phase: 40% 0.5 mol dm³ formic acid. Mobile phase: 15% ethyl acetate in hexane. Flow rate: 5 cm³ min⁻¹.

Fig. 5. (a) Cross-sectional view of the modified column inlet fitting. (a) ¼ inch×28 turns per inch – to accept standard LC flange fitting; (b) manifold (stainless steel); (c) collar (stainless steel); (d) piston (glass-filled teflon); (e) 'o'-rings; (f) annular feed passage (one of four); (g) sample feed passage; (h) porous teflon disc; (i) glass column (Jobling); (j) water jacket (Jobling); (k) column end-fitting cap (Jobling, drilled to accept manifold); (l) collet (Jobling, one of two); (m) column end fitting (Jobling). (b) Plumbing of the modified column inlet fitting and sample valve.

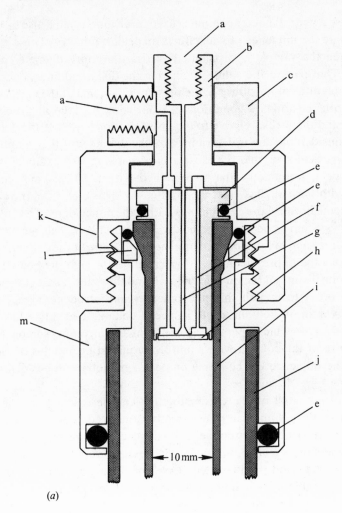

(a)

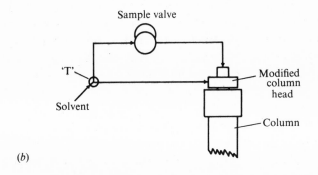

(b)

Because a wide diameter column is used, precautions must also be taken to eliminate the influence of wall effects on peak symmetry. These effects arise when the sample is distributed across the entire diameter of the column. That part of the solute occupying the disturbed flow zones near the wall migrates more slowly and elutes later than that at the axial centre where more uniform flow conditions prevail. The size of the disturbed flow zone around the walls, relative to the cross-sectional area of the column, can be considerable and is determined by the diameter of the column, the support particle size, and the closeness of packing. An example of the marked peak asymmetry that is obtained with a 10×450 mm column packed with Partisil 20 is illustrated in Fig. 4a. This effect can, however, be overcome because the column falls within the dimensions of an 'infinite diameter' system as defined by Knox & Parcher (1969). This means that, provided the sample is introduced into the axial centre of the column rather than across the entire diameter, it will migrate through the bed without entering the disturbed region near the wall and thus elute as a symmetrical peak. In order to exploit this phenomenon and introduce large sample volumes via an off-column valve, a special column inlet fitting has been designed (Fig. 5) which operates by confining the incoming sample to the axial centre of the column with a uniformly migrating annulus of mobile phase. The influence of the fitting on peak symmetry can be seen when comparing Figs. 4a and 4b.

The full potential of the chromatograph can be realised only if GAs eluting from the column can be accurately monitored. Since conventional HPLC detectors are not suitable, a homogeneous radioactivity monitor has been developed and this makes it possible rapidly to analyse an extract for [³H]- and [¹⁴C]-labelled GAs. The inclusion of this monitor in the system requires a suitable compromise to be made between chromatographic resolution and speed of analysis and detector sensitivity (Reeve & Crozier, 1977). This is achieved by selecting a suitable scintillant and scintillant:eluant ratio, matching the flow cell volume and geometry to the minimum chromatographic peak width and adjusting the overall flow rate to give an optimum value for flow cell transit time. When these parameters are optimised the monitor has a relative sensitivity of 3×10^3 dpm for ^{3}H and 1×10^3 dpm for ^{14}C for a solute eluting with a k' of 1.7. The monitor does not contribute to the total bandspreading of the chromatograph for solutes where $k' > 1.7$.

By manipulation of the hexane-ethyl acetate ratio a wide range of mobile phase solvent strengths can be used to provide rapid and effective separations. Fig. 6 illustrates the use of a gradient designed for the analysis of samples whose components span a wide range of polarities.

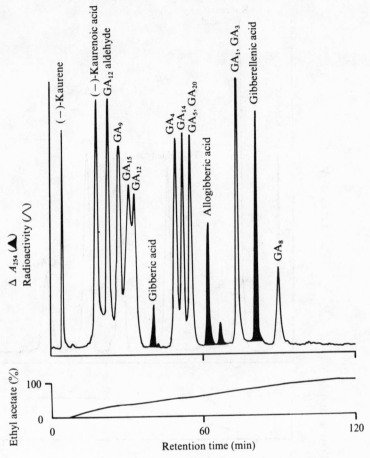

Fig. 6. Preparative HPLC of radioactive GAs and GA precursors with UV-absorbing internal markers. Column: 10×450 mm Partisil 20. Stationary phase: 40%, 0.5 mol dm^{-3} formic acid. Mobile phase: hexane-ethyl acetate gradient, as illustrated. Flow rate: 5.0 cm^3 min^{-1}. Sample: *ca* 24000 dpm $(-)$-$[^{14}C]$-kaurene; 50000 dpm $[^{14}C]GA_3$, $[^3H]GA_5$, $[^{14}C]GA_{12}$, $[^{14}C]GA_{15}$ and $[^3H]GA_{20}$; 100000 dpm $(-)$-$[^3H]$kaurenoic acid, $[^3H]GA_1$, $[^3H]GA_4$, $[^3H]GA_8$, $[^3H]GA_9$, $[^3H]GA_{12}$ aldehyde and $[^3H]GA_{14}$; and uncalibrated amounts of gibberic acid, allogibberic acid and gibberellenic acid. Detectors: radioactivity monitor 1800 cpm full scale deflection (FSD), UV monitor 0.5A FSD.

The GAs and GA precursors separate according to the degree of hydroxylation. Those compounds having no hydroxyl groups, such as $(-)$-kaurene, $(-)$-kaurenoic acid, GA_{12}-aldehyde, GA_9, GA_{15} and GA_{12}, elute first, followed by the monohydroxylated GAs (GA_4, GA_{14}, GA_5 and GA_{20}), the dihyroxylated compounds GA_1 and GA_3, and finally GA_8 which

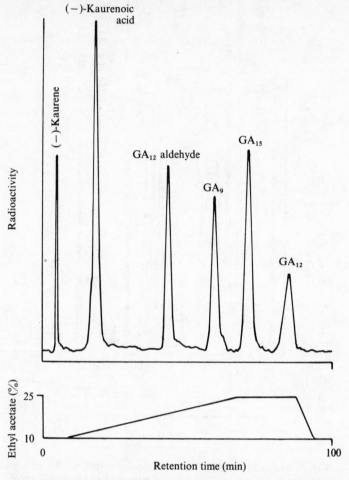

Fig. 7. Preparative HPLC of radioactive GAs and GA precursors using a restricted solvent gradient to give increased resolution at the non-polar end of the GA polarity spectrum. Column: 10×450 mm Partisil 20. Stationary phase: 40%, 0.5 mol dm^{-3} formic acid. Mobile phase: hexane-ethyl acetate gradient as illustrated. Flow rate: 5 cm^3 min^{-1}. Sample: *ca* 8000 dpm $(-)$-[^{14}C]kaurene; 20000 dpm [^{14}C]GA$_{12}$ and [^{14}C]GA$_{15}$ and 40000 dpm $(-)$-[^{3}H]kaurenoic acid, [^{3}H]GA$_9$ and [^{3}H]GA$_{12}$ aldehyde. Detector: radioactivity monitor, 600 cpm FSD.

has three hydroxyl groups. When wide range gradients of this type are employed, UV-absorbing markers are normally incorporated into the sample to allow precise determination of the relative retentions of the radioactive peaks. The UV trace in Fig. 6 shows the elution points of gibberic, allogibberic and gibberellenic acid, which are conveniently prepared by acid hydrolysis of GA$_3$ and are suitable internal markers.

In practice, it is not always necessary to cater for such a wide range

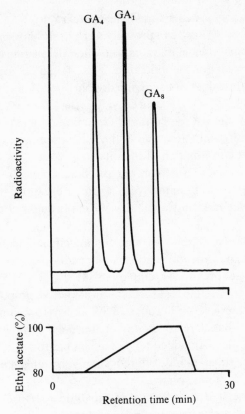

Fig. 8. Preparative HPLC of GA_1, GA_4 and GA_8 using a restricted
solvent gradient designed for rapid analysis. Column: 10×450 mm
Partisil 20. Stationary phase: 40%, 0.5 mol dm^{-3} formic acid. Mobile
phase: hexane–ethyl acetate gradient, as illustrated. Flow rate: 5.0
cm^3 min^{-1}. Sample: ca 20 000 dpm [^{3}H]GA_1, [^{3}H]GA_4 and [^{3}H]GA_8.
Detector: radioactivity monitor, 600 cpm FSD.

of sample polarity and a gradient operating over a more restricted range
of solvent strength can often be used. This is illustrated in Fig. 7 where
a 10–25% ethyl acetate gradient has been used to analyse the non-polar
end of the GA spectrum. The increased resolution has been achieved
through a more gradual and limited increase in the ethyl acetate concentration, which results in the generation of higher effective k' values.

When the metabolism of applied GAs involves successive hydroxy-
lations, the products are usually chromatographically distinct from each
other and from the precursor molecule. In such cases a considerable saving
can be made in the analysis time, because good separations are obtainable
without the need for high effective k' values. This is shown in Fig. 8 where
the solvent programme has been adjusted to allow the repeated separation
of GA_4, GA_1 and GA_8, at 30 min intervals.

'Analytical' high performance liquid chromatography

Whilst 'preparative' HPLC can provide much useful information on the identity of GA metabolites in extracts, retention data alone do not provide an unequivocal basis for identification. This should be apparent from the fact that the method does not separate double bond isomers such as GA_1 and GA_3, GA_4 and GA_7, and GA_5 and GA_{20}. It is therefore necessary to subject components of interest to additional chromatographic systems that will permit further differentiation, and at the same time increase the purity to such a level that confirmation of identity by mass spectrometry becomes possible. 'Analytical' HPLC on narrow bore columns with a 2–5 mm internal diameter is well suited to this role and can be readily applied because of the reduction in sample dry weight brought about by 'preparative' HPLC.

Although 'analytical' HPLC offers very high column efficiencies, it will only be fully effective if the procedure utilises a separatory mechanism quite different from those previously employed. HPLC on silica gel adsorption column is suitable in this regard and is an attractive proposition for the separation of isomeric GAs because of its ability to distinguish subtle differences in the spatial relationships of the polar groupings of structurally similar molecules. An added advantage is that the selectivity of the silica gel can be substantially altered simply by changing the reagent used to modify the mobile phase.

The main problem associated with the use of 'analytical' HPLC is the detection of the GAs. Although refractive index and far-UV monitors are often referred to as 'universal' detectors, they are not as useful as implied by the manufacturers' advertising literature which almost invariably fails to point out that they function only in a very restricted range of solvent conditions. Rather than be limited by the specialised needs of such detectors, it is more convenient to convert GAs to derivatives which absorb in an accessible region of the UV spectrum. As all GAs possess a carboxyl function at C-7 this can be achieved by esterification to produce GA benzyl esters (GABEs), which have a λ_{max} of 256 nm, and can be readily detected in a wide range of solvents with a standard UV monitor operating at 254 nm. The esterification step provides the additional advantage of eliminating the peak asymmetry which occurs when free carboxylic acids are chromatographed on silica gel adsorption columns.

We have developed a technique for the preparation of GABEs in microgram amounts using the reagent N,N'-dimethylformamide dibenzyl-acetal. This compound is prepared from N,N'-dimethylformamide dimethyl acetal (Pierce Methyl-8 concentrate, Pierce and Warriner (UK) Ltd, 44 Upper Northgate Street, Chester, Cheshire) by transesterification.

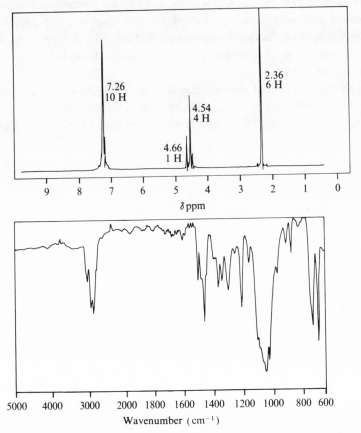

A 2:1 mole ratio mixture of dry benzyl alcohol and the dimethylformamide dimethylacetal is refluxed at 140 °C for 2 h, the methanol formed being removed during the course of the reaction by a stream of dry N₂. The reaction mixture is then purified by heating to 180 °C for 30 min at 5–20 torr to remove volatile components. The viscous, pale yellow oil that remains contains about 80% dimethylformamide dibenzylacetal which can be used without further purification. The purity of the product should be confirmed by nuclear magnetic resonance (NMR) and infra-red (IR) spectrometry and for reference purposes typical spectra are given in Fig. 9. The reagent is stable under dry storage conditions.

Fig. 9. 100 MHz nuclear magnetic resonance and KBr disc infra-red spectra of dimethylformamide dibenzylacetal.

The choice of solvent for the esterification reaction is critical. We have tested pyridine, dimethylformamide, dimethylsulphoxide (DMSO), ethyl acetate, acetone, diethyl ether, dichloromethane, dioxane and THF, and only the last two proved effective in promoting the benzyl esterification of GAs. Reactions carried out in THF can give rise to numerous products unless the solvent has been freshly distilled from lithium aluminium hydride. Dioxane is therefore more suitable provided it is purified by refluxing over sodium for four days. It can be stored in evacuated glass ampoules at −10 °C for at least one year.

After the extract has been dried over phosphorus pentoxide *in vacuo* for more than 6 h, the derivatisation is carried out in a septum capped 'Reactivial'. Dioxane, then dimethylformamide dibenzylacetal are added through the septum using 100 mm³ of each per milligram of sample. On heating to 70 °C monocarboxylic GAs, such as GA_1 and GA_3, are converted to their respective benzyl esters within 2 h. Dicarboxylic and tricarboxylic acids, such as GA_{14} and GA_{13}, require up to 4 h for complete esterification. Provided sufficient time is allowed for the reaction, the predominant product from GA_{14} is the *bis* benzyl ester (Fig. 10) and from GA_{13} the *tris* benzyl ester. For all classes of GAs, the yield of the respective benzyl ester is 80 to 85%.

At the completion of the reaction it is necessary to separate the GABE from the large excess of dimethylformamide dibenzylacetal. This is carried out by 'preparative' HPLC and is an informative step as the magnitude

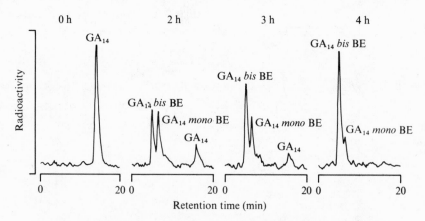

Fig. 10. A time-course study of the benzyl esterification of [³H]GA_{14} using preparative HPLC. Column: 10×450 mm Partisil 10. Stationary phase: 40%, 0.5 mol dm⁻³ formic acid. Mobile phase: 55% ethyl acetate in hexane. Flow rate: 5 cm³ min⁻¹. Sample: a dioxane–dimethylformamide dibenzylacetal-[³H]GA_{14} reaction mixture held at 70 °C. Aliquots of *ca* 15000 dpm analysed 0, 2, 3 and 4 h after the start of the reaction. Detector: radioactivity monitor 200 cpm FSD

of the change in retention of the GA on esterification gives an indication of the number of benzyl groups contained by the molecule. Furthermore, as not all components in an extract will esterify and thus change retention, a substantial increase in the sample purity is obtained. These methods have been used to produce the benzyl esters of GA_1, GA_3, GA_4, GA_5, GA_7, GA_9, GA_{13}, GA_{14}, GA_{15} and GA_{20}, all of which have been characterised by NMR, IR, UV and mass spectrometry. The derivatisation products of GA_8 appear, however, to include a formyl addition product of GA_8BE, although this has not as yet been characterised.

The chromatograph used for the separation and purification of GABEs is constructed from conventional 'analytical' HPLC instrumentation and is illustrated in Fig. 11. Mobile phase is delivered by a constant pressure, pneumatic displacement pump to a prepacked 4.6×500 mm Partisil 10 column. Samples of up to 10 mm³ in volume are injected 'stop flow' and 'on column' by means of a microlitre syringe. UV-absorbing components eluting from the column are detected by an UV monitor operating at 254 nm and fitted with a 10 mm³ flow cell. Solvent may either be collected

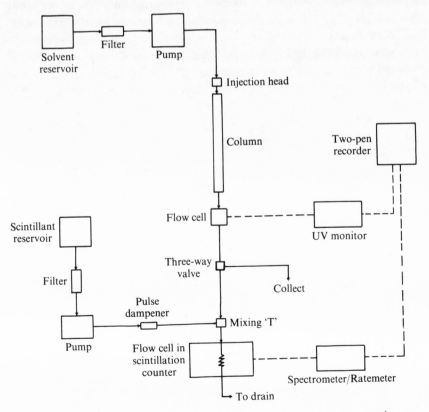

Fig. 11. The analytical HPLC with UV and radioactivity monitors.

at the exit port of the UV flow cell or directed to an homogeneous radioactivity monitor similar to that used with the preparative HPLC. A flow rate of 1.0 cm³ min⁻¹ gives a suitable compromise between the speed of analysis and radioactivity monitor sensitivity. This corresponds to a solvent linear velocity of 1.3 mm s⁻¹ at which the column generates up to 8000 theoretical plates with a speed of analysis of 3.2 effective plates per second, for a solute eluted with a k' value of 2.0 by a dichloromethane carrier. This normally requires a column inlet pressure of 400 psi. Under these conditions the UV monitor has a minimum detection limit of ca 0.5 μg for GA *mono* benzyl esters. With *bis* and *tris* derivatives the figure is correspondingly lower. The radioactivity monitor will detect 6×10^3 dpm of ^3H and 2×10^3 dpm of ^{14}C, and exhibits negligible loss of chromatographic resolution at k' values > 1.5.

The mobile phase employed for silica gel adsorption HPLC has two funtional components: the carrier and the modifier. The major component, the carrier, normally consists of one or two solvents exhibiting relatively low strength towards the compounds being chromatographed, and in order to decrease retention times is modified with small amounts of an agent such as THF, DMSO, acetonitrile or ethanol. The modifiers interact with the silica surface and thereby subtly alter the steric and electronic environment of the adsorption site. Hence for any given carrier a range of selectivities may be achieved through the use of various modifiers.

The nature of the carrier has gross control over the chromatographic process, and its selection is best guided by an understanding of the factors contributing to the retention behaviour of GABEs on silica adsorption systems. Increasing the degree of hydroxylation of a GABE markedly increases retention values, whilst increasing the number of benzyl groups reverses this effect. This is illustrated in Table 1. In addition, the retention of any one GABE is predominantly controlled by the degree of nucleo-

Table 1. *The relationship between the number of hydroxyl functions and benzyl groups and the retention behaviour of GABEs on silica gel adsorption HPLC columns*

	Elution order of benzyl esters				
	$GA_{13} \approx GA_{14} \approx GA_9, GA_{15}$	>	GA_4, GA_7 GA_5, GA_{20}	>	GA_1, GA_3
Hydroxyl groups	2	1	0	1	2
Benzyl groups	3	2	1	1	1

Table 2. *The influence of two parameters of solvent strength on the retention behaviour of GABEs on silica gel adsorption HPLC columns*[a]

Solvent[b]	δ^o	δ^a	Capacity factor (k')		
			GA₃BE	GA₄BE	GA₉BE
Hexane	0	0	Very large	Very large	Very large
Dichloromethane	5.5	0.5	> 10	> 10	1.9
Diethyl ether	2.0	2.0	1.9	0.2	Small
Ethyl acetate	3.0	2.0	0.3	Small	Small
THF	4.0	3.0	Very small	Very small	Very small
DMSO	7.5	5.0	Very small	Very small	Very small

[a] The solvent polarity, δ^o, and nucleophilicity, δ^a, are two of a number of solvent strength parameters used by Snyder (1971) to help to rationalise the influence of solvent characteristics in liquid chromatography. δ^o gives a measure of the solvents' ability to take part in dipole–dipole interactions, while δ^a indicates the potential for weak bond formation by donation of electrons.
[b] Arrow indicates increasing solvent strength.

philic character displayed by the carrier. This effect can be seen in Table 2 where the eluotropic solvent series for GABEs runs hexane → dichloromethane → diethyl ether → ethyl acetate → THF → DMSO, in order of increasing solvent nucleophilicity, rather than the more usual series, based on increasing dipole moment, hexane → diethyl ether → ethyl acetate → THF → dichloromethane → DMSO. In general, dichloromethane is an excellent carrier for the separation of GABEs. It combines the advantages of low viscosity, and UV transparency, with the fact that its solvent strength is in the main derived from its high polarity (δ^o) (see Table 2). This means it can be modified by the addition of small amounts of nucleophilic agents such as THF and DMSO to give a range of mobile phases highly selective for GABEs. In order fully to exploit the selectivity of the mobile phase it is important that the carrier strength be such that no more than 10% (v/v) modifier is required to obtain practical retention values. For example, although GA₄BE is eluted with a k' of 1–2 by both 3% THF in dichloromethane and 25% THF in hexane, only the mobile phase using the lower concentration of modifier is capable of separating the GA₄BE from its $\Delta^{1,2}$ analogue, GA₇BE.

When a 1,2 double bond is introduced into the *A* ring of a GABE the 3β-hydroxyl group changes orientation with respect to both the lactone and carboxyl functions, and it is probably this difference in orientation that provides the basis for the separation of GA₄BE and GA₇BE (relative

Table 3. *The influence of solvent polarity, δ^o, nucleophilicity (δ^a) and dispersive index (δ^d) on the relative retention (α) of $\Delta^{2,3}$- and $\Delta^{1,2}$-GABE analogues*

				Relative retention (α)	
Solvent	δ^o	δ^a	δ^d	GA$_{4/7}$BE	GA$_{5/20}$BE
4% THF in dichloromethane	5.5	0.6	6.4	1.15	< 1.1
5% THF in dichloroethane	4.0	0.2	8.2	1.20	1.15

retention, $\alpha = 1.13$) and GA$_1$BE and GA$_3$BE ($\alpha = 1.15$) in THF-modified dichloromethane. In keeping with this suggestion, the $\Delta^{2,3}$ analogues GA$_5$BE and GA$_{20}$BE, which contain no hydroxyl groups at C-3, cannot be resolved. Other factors do, however, contribute to this separation, as can be demonstrated by the substitution of dichloroethane for dichloromethane. The only major difference between these two solvents is that the former has a higher dispersive power which tends to emphasise differences in the degree of saturation of GABE analogues. In Table 3 it can be seen that increasing the dispersive power of the solvent facilitates the separation of GA$_5$BE and GA$_{20}$BE and further resolves GA$_4$BE and GA$_7$BE. Separation of these compounds can also be achieved through the use of DMSO-modified dichloromethane, which gives α values of 1.17 and 1.42 for the pairs GA$_{5/20}$BE and GA$_{4/7}$BE respectively. The nature of the separation achieved is quite different to that occurring in THF-modified solvents and results in the complete reversal of the elution order with respect both to location of the hydroxyl group and to the degree of unsaturation. These differences, which can be of considerable value in separation and purification of isomeric GABEs, are illustrated in Fig. 12.

The efficiency and selectivity of silica gel adsorption HPLC are obtained at the expense of sample recovery. However, by maintaining the water content of the mobile phase at approximately half the saturation value, it is possible to obtain recoveries of 60% and concomitantly to increase the peak symmetry and column efficiency. The precise water content is not critical as it has little influence on the retention behaviour of GABEs in the solvent systems described.

In the case of [^{3}H]GA metabolites from *Phaseolus* seedling extracts that have been purified by GPC, charcoal adsorption chromatography, 'preparative' HPLC and further 'preparative' HPLC and 'analytical' HPLC as benzyl esters, the purity of the compounds of interest is almost

Mobile phase:
Hexane-dichloromethane-DMSO
(50:50:1)

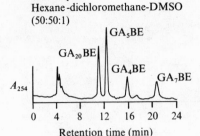

Mobile phase:
Hexane-dichloromethane-DMSO
(25:75:1)

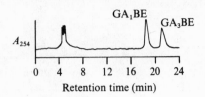

Mobile phase:
Dichloromethane-THF (97:3)

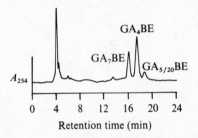

Mobile phase:
Dichloromethane-THF (92·8)

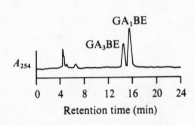

Fig. 12. The influence of DMSO and THF modifiers on the analytical HPLC retention characteristics of some GABEs. Column: 4.6×500 mm Partisil 10. Mobile phase: as indicated on figure. Flow rate: 1.6 cm³ min⁻¹. Sample: GA₄BE, GA₅BE, GA₇BE and GA₂₀BE or GA₁BE and GA₃BE. Detector: UV monitor. Ratios are all by volume.

always well above 80%. The main contaminants present at this stage originate in the HPLC solvents. Although these solvents are carefully purified, the weight of the extract is often as low as 1 μg and unavoidable contamination from the laboratory environment invariably occurs. It is possible to effect the final purification by introducing the sample into the mass spectrometer via a gas chromatograph interface. However, we find it simpler to carry out a final 'analytical' HPLC step and introduce the sample into the mass spectrometer by a direct insertion probe. This chromatographic step is designed to reduce contamination by eluting the compound of interest as a very narrow band with a volume of no more than 200–250 mm³. A 4.6×250 mm prepacked Partisil 10 column fitted with a septumless, on-column injection port is eluted with a mobile phase of freshly distilled diethyl ether. The UV-absorbing peak representing the GABE is collected in a microcentrifuge tube previously cleaned by flaming to red heat. The diethyl ether is evaporated off under a stream of filtered, dry nitrogen and the sample taken up in two 10 mm³ volumes of dry, distilled acetone, transferred to the capillary of the direct insertion probe, dried and introduced into the mass spectrometer.

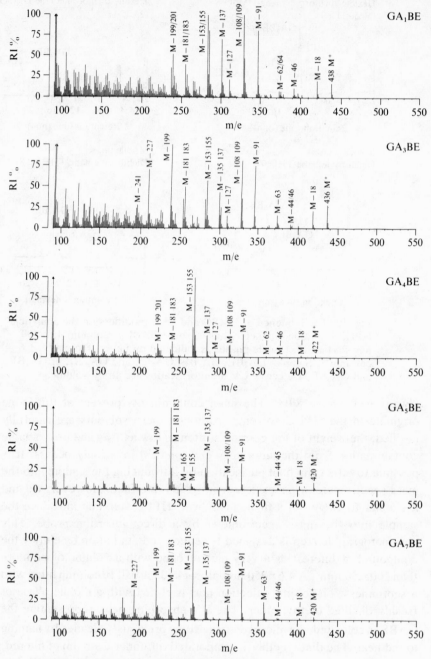

Fig. 13. Caption is at the foot of page 66.

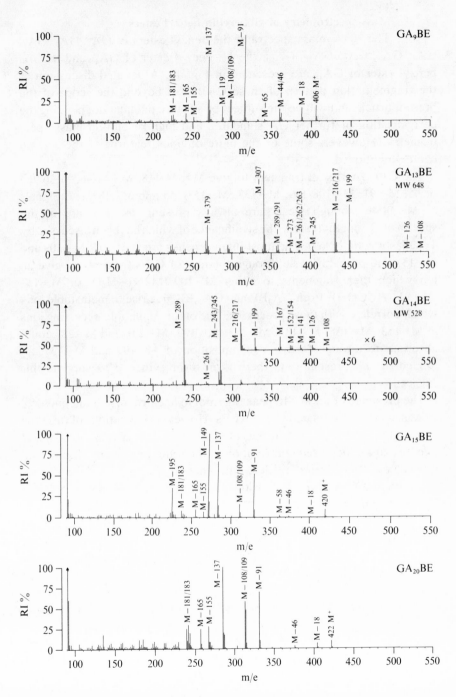

Mass spectrometry of gibberellin benzyl esters

The 70 eV mass spectra of the benzyl esters of GA_1, GA_3, GA_4, GA_5, GA_7, GA_9, GA_{15}, GA_{20}, the *bis* benzyl ester of GA_{14} and the *tris* benzyl ester of GA_{13} are presented in Fig. 13. A detailed discussion of the fragmentation pattern of these GABEs is beyond the scope of the present article and will be the subject of a future publication, pending the derivatisation of further GA standards and high resolution mass spectrometry. However, some of the more obvious features of the spectra can be mentioned.

The C19 GABEs all fragment to give M^+, $M-18$, $M-44/46$, $M-62/64$ or 65, $M-91$, $M-108/109$, $M-137$, $M-153/155$ and $M-181/183$. Signals at $M-44/46$, though weak, are always present, as they are in the spectra of GA methyl esters, the significance of which has been discussed by Takahashi *et al.* (1969). The most intense peaks are at $M-91$, $M-109$ and $M-137$ and probably arise through the loss of C_7H_7 from M^+ to give an ion which then fragments to either $M-109$ ($M-91-H_2O$) or $M-137$ ($M-91-HCOOH$). Both GA_3BE and $GA_{20}BE$ give rise to metastable ions which correlate with these processes. The other commonly occurring ions at $M-153$, $M-155$, $M-181$, $M-183$, $M-199$, $M-201$ and $M-227$ could all originate through further decomposition of $M-109$ and $M-137$, although at the present time there is no firm evidence to support this suggestion.

The presence of an $M-18$ peak does not in itself represent unequivocal evidence for hydroxylated C19 GABEs. However, indications of this type of structure can be found in the intensity of the tropylium ion signal at m/e 91 and an increased contribution to the total ion current of ions with m/e values < 300. There are clear differences in the spectra of the various hydroxylated C19 GABEs. Substitution at C-3 reduces the intensity of $M-137$ and markedly increases the value of $M-153$, $M-181$ and $M-199$. A peak at $M-127$ is also observable. On the other hand, although hydroxylation at C-13 has little effect on $M-137$ it does strengthen the signal at $M-155$, $M-165$ and $M-181/183$. The relative height of the peaks at $M-153$ and $M-155$ are of interest (1) as they are roughly equal in size for 3,13-dihydroxy GABEs, (2) as $M-153$ predominates with 3-hydroxy GABEs, while (3) in 13-hydroxy GABEs $M-155$ is the major contributor.

Fig. 13. Mass spectra of GABEs. Samples introduced into an AEI MS 30 mass spectrometer by direct probe. Source temperature: 230 °C; electron beam energy: 70 eV; and spectra recorded at a scan speed of 30 s decade^{-1}. RI = relative intensity. For $GA_{14}BE$ the inset has a six-fold increase in ordinate scale sensitivity (i.e. magnification).

Other features of the C19 GABEs which may prove to be characteristic are the ions at $M-119$ for GA_9BE and the increase in intensity of $M-135$ for $\Delta^{1,2}$ and $\Delta^{2,3}$ GABEs. The peak at $M-63$, which is a feature of $\Delta^{1,2}$ 3-hydroxy GABEs also shows up in the spectra of both GA_3 and GA_7 methyl esters, and has been commented on by Binks *et al.* (1969). In the case of unknown GABEs that exhibit some of the spectral characteristics of hydroxy C19 GAs, yet do not correspond to any of the available reference spectra, it may be necessary to resort to trimethylsilylation or acetylation in order to confirm the degree of hydroxylation of the molecule.

The only C20 GABEs that have been prepared to date are those of GA_{13}, GA_{14} and GA_{15}. The spectrum of $GA_{15}BE$ shows most of the characteristics displayed by its C19 analogue GA_9BE, except that the $M-18$ peak is very weak and $M-119$ is absent. The occurrence of an intense peak at $M-149$ and one of lower intensity at $M-195$ renders the $GA_{15}BE$ spectrum distinctly different from the spectra of the other GABEs so far examined.

The spectrum of the *bis* benzyl ester of GA_{14} consists essentially of only two major peaks, namely those at $M-243$ and $M-289$. A molecular ion cannot be observed and the first relatively stable fragment is $M-108$. This appears to decompose in a manner analogous to that of typical C19 GAs to give $M-126(-108-18)$, $M-141(-108-33)$, $M-152/154(-108-44/46)$, $M-167(-108-59)$, $M-199(-108-91)$, $M-216/217(-108-108/109)$, $M-243/245(-108-135/137)$, $M-261(-108-153)$ and $M-289(-108-181)$. The $M-108$ could be produced by the facile loss of the C-19 benzyl group as C_7H_7OH to give an ion similar in many respects to the molecular ion of the C19 GABEs.

The spectrum of the *tris* benzyl ester of GA_{13} also lacks a molecular ion. However a weak signal corresponding to $M-108$ can be detected and is presumably the result of the loss of C_7H_7OH from the benzyl ester function at C-19. Intense metastable peaks provide some evidence for the sequential decomposition of this ion by the loss of $C_7H_7(M-199)$ and then $C_7H_7OH(M-307)$. As in the case of the *bis* benzyl ester of GA_{14}, most of the minor peaks in the *tris* benzyl ester of GA_{13} spectrum can be rationalised by considering them to be the result of the decomposition of the $M-108$ ion in a manner analogous to the events occurring with C19 GABEs. Thus $M-108$ could give rise to $M-126(-108-18)$, $M-154$ $(-108-46)$, $M-199(-108-91)$, $M-216/217(-108-108/109)$, $M-245$ $(-108-137)$, $M-261/262/263(-108-153/154/155)$, $M-273(-108-165)$, $M-289/291(-108-181/183)$. On this basis it is likely that the C-7 benzyl ester is the second benzyl moiety to be expelled from the molecule, and this yields $M-108-91$, which decomposes to give an ion series analogous to that produced by C19 GABEs. The initial processes in this sequence, that is M−108−109 and M−108−137 can be partially validated by correlation with metastable peaks. However, M−108−91 can also decompose by expulsion of the third benzyl function at C-20 as C_7H_7OH to give M−289(M−108−91−108) and an intense metastable ion provides evidence for such a transition.

Until additional *bis* and *tris* benzyl esters of GAs are prepared it is difficult to assess the potential diagnostic value of the mass spectra of these compounds, although the lack of a molecular ion and the paucity of higher mass fragments suggest they are likely to be of limited value. More information may be obtained by using an ether derivative such as the trimethylsilyl ether of the GABEs. Certainly the mass spectra of the trimethylsilyl ethers of GA methyl esters yield more information than the corresponding GA methyl esters (Binks *et al.*, 1969).

Analysis of radioactive gibberellin metabolites

The combination of group separatory procedures and HPLC techniques that have been described provide an effective means for the purification and subsequent characterisation of GA metabolites formed during radioactive tracer studies. Although the procedures used must be guided to some extent by the nature of the impurities in the individual extracts, a specific example is given below to illustrate one of the ways in which the techniques can be employed. In this instance a [³H]GA₉ metabolite, which was originally present as a minor component in an acidic, ethyl acetate-soluble extract, was purified to the point where characterisation by direct-probe mass spectrometry became possible.

Two hundred light-grown *Phaseolus coccineus* seedlings were treated with 20 μCi of [2,3-³H]GA₉ (47 ci mmol⁻¹) diluted by the addition of 3 mg of carrier. Eight hours later, the plants (1.4 kg fresh weight) were harvested and after extraction and partitioning an acidic, ethyl acetate-soluble fraction (1.0 g, 18×10^6 dpm) was obtained. This was subjected to GPC and charcoal adsorption chromatography to produce a semi-purified extract (100 mg, 12×10^6 dpm) which on analysis by 'preparative' HPLC was shown to contain two peaks of radioactivity, one corresponding to the GA₉ substrate and the other to GA₅ and GA₂₀ (Fig. 14). The metabolite peak contained 4.8×10^6 dpm of radioactivity and weighed 5.0 mg. A one-quarter aliquot was benzyl esterified by treatment with N,N'-dimethylformamide dibenzylacetal. Examination of the reaction mixture by 'preparative' HPLC revealed one peak of radioactivity (1.1×10^6 dpm, < 1 mg) that eluted at the same point as GA₅BE and GA₂₀BE. The reduced sample

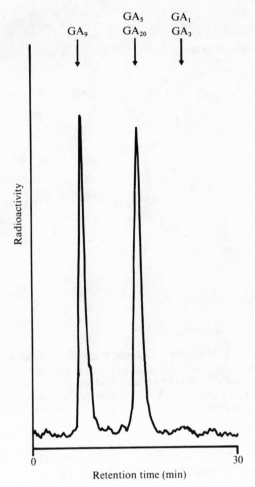

Fig. 14. Preparative HPLC of a semi-purified, acidic, ethyl acetate-soluble fraction from 200 light-grown *Phaseolus coccineus* seedlings extracted 8 h after the application of [^{3}H]GA$_9$ (3 mg, 44×10^6 dpm). Column: 10×450 mm Partisil 10. Stationary phase: 40%, 0.5 mol dm^{-3} formic acid. Mobile phase: 60–80% ethyl acetate in hexane. Flow rate: 5 cm^3 min^{-1}. Sample: 102 mg, 12.1×10^6 dpm of extract. Detector: radioactivity monitor 1800 cpm FSD. Splitter: 0.6% to radioactivity monitor, 99.4% to hold-up coil. Arrows indicate elution points of GA$_1$, GA$_3$, GA$_5$, GA$_9$ and GA$_{20}$.

weight meant that further purification could be carried out by 'analytical' HPLC. The use of a DMSO-modified mobile phase indicated the presence of a single zone of radioactivity (0.77×10^6 dpm) that chromatographed with a large UV-absorbing impurity peak and had a retention time equivalent to that of GA$_{20}$BE but not GA$_5$BE (Fig. 15a). The presumptive

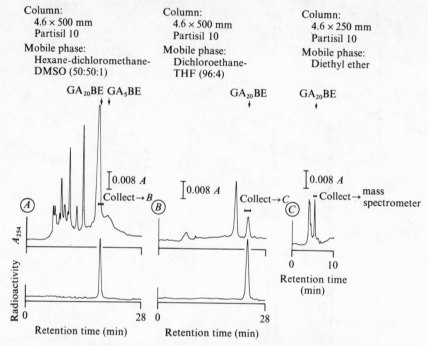

Fig. 15. Analytical HPLC of GABEs. Column and mobile phase as indicated. Flow rate: 1 cm³ min⁻¹. Sample: (A) benzyl ester of [³H]GA₉ metabolite – see Fig. 14; (B) eluate collected from A; (C) eluate collected from B. Detector: UV and radioactivity monitors. Arrows indicate elution points of GA₅BE and GA₂₀BE. Ratios are all by volume.

benzyl ester of the [³H]GA₉ metabolite (0.37×10^6 dpm) separated from the impurity in a THF-modified solvent (Fig. 15b) and was further chromatographed in a diethyl ether mobile phase to reduce contamination from solvent impurities (Fig. 15c). The UV peak corresponding to GA₂₀BE was examined by direct-probe mass spectrometry and the mass spectrum obtained proved to be identical to that of GA₂₀BE (Fig. 16).

The extinction coefficient of GA₂₀BE at 254 nm is 205 l mol⁻¹ cm⁻¹. The UV peak corresponding to GA₂₀BE in Fig. 15b is therefore equivalent to 26 μg which is in agreement with an estimate of 25 μg calculated on the basis of the recoverable radioactivity. This close correlation eliminates the possibility of significant dilution of the label and it can be concluded that the GA₂₀ mass spectrum is that of a metabolite formed from [³H]GA₉ and that no contribution has been made by endogenous GA₂₀.

The mass spectrum in Fig. 16 indicates that the GA₂₀ formed from [³H]GA₉ was essentially pure. 'Preparative' and 'analytical' HPLC there-

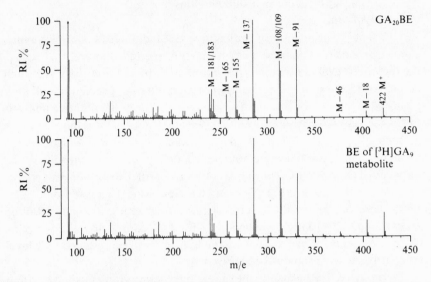

Fig. 16. Mass spectrum of authentic GA$_{20}$BE and the purified benzyl ester of [^{3}H]GA$_9$ metabolite. RI = refractive index.

fore brought about a 310-fold improvement in the purity of the semi-purified, acidic, ethyl acetate soluble extract. In spite of the number of chromatographic steps involved, 18% of the GA$_{20}$ originally present was recovered and this was associated with a 1700-fold reduction in sample weight. In practical terms, the HPLC procedures enabled direct-probe mass spectrometry to be used to obtain the mass spectrum of a compound present in an extract at the 0.3% level. Depending upon the nature of the impurities present, extracts of substantially lower purity can be similarly analysed. For example, an extract containing a 0.03% concentration of a metabolite of [^{3}H]GA$_4$ was reduced in weight by a factor of 10000 to give a compound that was readily identified as GA$_1$BE by mass spectrometry (Crozier & Reeve, 1977). Through the use of HPLC it should in theory be possible to characterise, by direct-probe mass spectrometry, radioactive GA metabolites that constitute as little as 0.005% of an extract. Any further refinement will require the use of GC–MS with computerised signal processing facilities (see Gaskin & MacMillan, this volume). By combining the described HPLC methodology with existing GC–MS technology it should be possible to analyse extracts containing as little as 0.0001% GA.

Analysis of endogenous gibberellins
Qualitative

The identification of radioactive GA metabolites is often a much simpler proposition than the characterisation of endogenous GAs, because the radioactive label facilitates detection, and, in addition, pool sizes can be increased through the application of high doses of precursor and other forms of manipulation such as the use of a cold trap. Qualitative analysis of GAs in vegetative tissues by GC–MS usually necessitates the extraction of tens of kilograms of plant material and the high dry weight of the resultant acidic, ethyl acetate-soluble fraction may tax the capacity of the initial purification steps. The use of countercurrent distribution procedures (Crozier *et al.*, 1969, 1971; Bowen, Crozier, MacMillan & Reid, 1973) must therefore be seriously considered along with polyamide (Railton & Wareing, 1973) GPC, Sephadex G-10, PVP and charcoal adsorption chromatography, if the weight of the extract is to be reduced to a level compatible with 'preparative' HPLC.

'Preparative' HPLC using a hexane-ethyl acetate gradient similar to that illustrated in Fig. 6 provides a good separation of a wide spectrum of GAs. A minimum of 200 fractions must be collected if the peak capacity of the system is to be fully utilised. The analysis of these fractions represents a major problem because of the lack of a specific label, such as radio-activity, that would permit the rapid detection of the GAs. One method is to methylate and trimethylsilylate and to examine each fraction by GC–MS. This is, however, not only extremely time consuming but because of a lack of purity many of the GAs may go undetected. In most instances it would be more practical to locate the GA-containing fractions by bioassay, and to accept as unavoidable the loss of biologically inactive components. Subsequent purification of the active zones should avoid the formation of biologically inactive GA derivatives until the purity requirements of GC–MS can be met. In this situation, therefore, 'analytical' HPLC of the GAs as their benzyl esters is of no value as a purification tool. It is, however, feasible to purify the GAs as free acids on 'analytical' HPLC silica gel adsorption systems provided an acidified mobile phase is used. Thin-layer chromatography solvent systems for GAs (Kagawa, Fukinbara & Sumiki, 1963; Cavell, MacMillan, Pryce & Sheppard, 1967) can readily be adapted for this purpose. Zones of biological activity eluted from the 'analytical' HPLC column can, if necessary, be further purified by altering the selectivity of the column. Once adequate sample purity is achieved the extracts can be subjected to methylation, silylation and GC–MS. An alternative procedure for the purification of the biologically active zones from the 'preparative' HPLC column prior to GC–MS involves the for-

mation of trimethylsilyl derivatives which can be analysed by preparative
gas liquid chromatography. GA trimethylsilyl esters and ethers hydrolyse
to release the free acid, so bioassays can be used to detect GA-like
activity in fractions collected from the gas chromatograph (Lorenzi,
Horgan & Heald, 1975, 1976).

The problems encountered in the qualitative analysis of endogenous
GAs will clearly differ greatly from one tissue to another and the methodology outlined here is no more than a guide to some of the options that
are available. The exact strategy employed, especially in the later stages
of the analysis, is best determined by an on the spot assessment rather
than the application of hard and fast rules.

Quantitative

Much use has been made of bioassays to give 'quasi'-quantitative
estimates of endogenous GA levels in plant tissues. In spite of recent
advances in analytical instrumentation, the bioassay remains an effective
means of detecting very low concentrations of GAs in plant extracts,
provided a markedly positive rather than barely significant response is
obtained. There are, however, other limitations that should be noted. The
quantitative accuracy of bioassay-based estimates is heavily dependent
upon the purity of the extract, as the bioassay response reflects the interaction between GAs and inhibitory compounds present in the sample.
Thus extensive purification is necessary to reduce the contribution of
inhibitory compounds which would otherwise mask to some degree the
presence of any GAs. This can usually be achieved by subjecting the
acidic, ethyl acetate-soluble fraction to GPC and charcoal adsorption
chromatography, or other group separatory procedures, before separating
the GAs by 'preparative' HPLC. Once again it is necessary to collect at
least 200 fractions if the degree of overlap between inhibitors and zones
of biological activity is to be reduced to a level that will permit meaningful
quantification. As an alternative to 'preparative' HPLC, the GAs in the
semi-purified extract could be separated by thin-layer chromatography.
While this would necessitate bioassaying only 10–20 fractions, it is highly
doubtful that the convenience gained would justify the greatly reduced
reliability of the results. If a 'preparative' HPLC system is not available,
less drastic compromises can be made. Classical liquid chromatography
using either a dextran (MacMillan & Wels, 1973) or a silica gel (Powell
& Tautvydas, 1967) support offers results far superior to those from
thin-layer chromatography and does not require the use of expensive
equipment.

The reliability of quantitative estimates of GA levels is further compounded by the fact that individual GAs exhibit widely varying biological potencies. It can therefore be very misleading to express levels of unknown GA-like compounds in micrograms of GA_3 equivalents, as the actual amount present may differ by up to several orders of magnitude. If previous work has shown that a particular zone of activity is exclusively associated with a certain GA, in subsequent experiments it will be possible to make allowances for the dose–response curve of that GA rather than GA_3, and thereby greatly increase the accuracy of quantitative estimates. The differing response spectrum of each GA bioassay, and their variation in susceptibility to the inhibitors and toxic materials found in typical plant extracts, make the choice of bioassay an important consideration. Our experience suggests that a combination of the barley aleurone, 'Tanginbozu' dwarf rice and lettuce hypocotyl assays provides good results (Reeve & Crozier, 1975).

Although bioassays can be of considerable diagnostic value they leave much to be desired as an analytical technique. Physiochemical procedures such as mass fragmentography, which employs a mass spectrometer as a selective ionisation detector for a gas chromatograph by monitoring only at those m/e values that are pertinent to the compound being analysed, are potentially much more precise (Holmstedt & Lindgren, 1974; Palmer & Holmstedt, 1975). At the present time the application of mass fragmentography to the quantitative analysis of GAs and other plant hormones is very much in its infancy although it has been used to determine the levels of GA_4 and GA_9 in wheat chloroplasts (Browning & Saunders, 1977) and indole-3-acetic acid and abscisic acid in *Zea mays* roots (Rivier & Pilet, 1974; Rivier, Milon & Pilet, 1977). The procedure is also reviewed by Gaskin & MacMillan (this volume). The main problem lies not so much in obtaining the necessary sensitivity and precision as in achieving sufficient selectivity to guarantee that the detector response is exclusive to the compound under study. When extracts have been extensively purified, more reliance can be placed upon quantitative estimates obtained by mass fragmentography, because of the reduced risk of interference by ions derived from spurious molecules. The 'preparative' and 'analytical' HPLC systems used to analyse [³H]GA metabolites can increase sample purity by many orders of magnitude while maintaining high GA recovery levels and they are therefore potentially well suited to the purification of endogenous GAs prior to mass fragmentography. Quantitative accuracy can be enhanced by using deuterium-labelled GAs as internal markers so that it becomes possible to compensate for losses occurring during sample purification as well as short-term variations in the sensitivity of the mass spectrometer.

We would like to thank Dr Linda Nash who generously allowed us to use some of her unpublished data, Mr W. McNulty for the skilful construction of the modified column inlet fitting illustrated in Fig. 5, and Professor R. P. Pharis, Professor J. E. Graebe and Dr J. MacMillan for generous gifts of radioactive GAs. Much of the work reported in this publication was supported by a Science Research Council (UK) Grant to AC. The Royal Society provided funds for the purchase of an on-stream radio-activity monitor.

References
The following four books are recommended reading as an introduction to the practice and elementary theory of HPLC.

Hadden, N., Baumann, F., MacDonald, F., Munk, M., Stevenson, R., Gere, D., Zamaroni, F. & Majors, R. (1971). *Basic Liquid Chromatography*. Varian Aerograph Publication.

Kirkland, J. J. (1971). *Modern Practice of Liquid Chromatography*. New York: Wiley-Interscience.

Parris, N. A. (1976). Instrumental Liquid Chromatography. *Journal of Chromatography Library*, vol. 5. Amsterdam: Elsevier.

Snyder, L. R. & Kirkland, J. J. (1974). *Introduction to Modern Liquid Chromatography*. New York: Wiley-Interscience.

Binks, R., MacMillan, J. & Pryce, R. J. (1969). Combined gas chromatography-mass spectrometry of the methyl esters of gibberellins A_1 to A_{24} and their trimethylsilyl ethers. *Phytochemistry*, 8, 271–84.

Bowen, D. H., Crozier, A., MacMillan, J. & Reid, D. M. (1973). Characterization of gibberellins from light-grown *Phaseolus coccineus* seedlings by combined GC-MS. Phytochemistry 12, 2935–41.

Browning, G. & Saunders, P. F. (1977). Membrane localised gibberellins A_9 and A_4 in wheat chloroplasts. *Nature, London*, 265, 375–7.

Cavell, B. D., MacMillan, J., Pryce, R. J. & Sheppard, A. C. (1967). Thin-layer and gas-liquid chromatography of the gibberellins; direct identification of the gibberellins in a crude plant extract by gas–liquid chromatography. *Phytochemistry*, 6, 867–74.

Crozier, A., Aoki, H. & Pharis, R. P. (1969). Efficiency of counter-current distribution, Sephadex G-10 and silicic acid partition chromatography in the purification and separation of gibberellin-like substances from plant tissue. *Journal of Experimental Botany*, 20, 786–95.

Crozier, A., Bowen, D. H., MacMillan, J., Reid, D. M. & Most, B. H. (1971). Characterization of gibberellins from dark-grown *Phaseolus coccineus* seedlings by gas–liquid chromatography and combined gas chromatography–mass spectrometry. *Planta, Berlin*, 97, 142–54.

Crozier, A. & Reeve, D. R. (1977). The application of high performance liquid chromatography to the analysis of plant hormones. In *Plant Growth Regulation*, ed. P. E. Pilet, pp. 67–76. Berlin: Springer.

Done, J. N., Kennedy, G. J. & Knox, J. H. (1972). Revolution in liquid chromatography. *Nature, London*, **237**, 77–81.

Done, J. N., Knox, J. H. & Loheac, J. (1974). *Applications of High-speed Liquid Chromatography*, 1st edn. London: John Wiley & Sons.

Durley, R. C. & Pharis, R. P. (1972). Partition coefficients of 27 gibberellins. *Phytochemistry*, **11**, 317–26.

Glenn, J. L., Kuo, C. C., Durley, R. C. & Pharis, R. P. (1972). Use of insoluble polyvinylpyrrolidone for purification of plant extracts and chromatography of plant hormones. *Phytochemistry*, **11**, 345–51.

Holmstedt, B. & Lindgren, J. E. (1974). Use of gas chromatography–mass spectrometry in toxicological analysis. In *The Poisoned Patient: The Role of the Laboratory*, ed. R. Porter & M. O'Connor, Ciba Foundation Symposium 26 (NS), pp. 105–24. Amsterdam: Elsevier.

Kagawa, T., Fukinbara, T. & Sumiki, Y. (1963). Thin layer chromatography of gibberellins. *Agricultural and Biological Chemistry*, **27**, 598–9.

Knox, J. H. (1974). High performance liquid chromatography. *Laboratory Equipment Digest*, **12**, 51–64.

Knox, J. H. & Parcher, J. F. (1969). Effect of column to particle diameter ratio on the dispersion of unsorbed solutes in chromatography. *Analytical Chemistry*, **41**, 1599–606.

Lorenzi, R., Horgan, R. & Heald, J. K. (1975). Gibberellins in *Picea sitchensis* Carriere: seasonal variation and partial characterization. *Planta, Berlin*, **126**, 75–82.

Lorenzi, R., Horgan, R. & Heald, J. K. (1976). Gibberellin A$_9$ glucosyl ester in needles of *Picea sitchensis*. *Phytochemistry*, **15**, 789–90.

MacMillan, J. (1972). A system for the characterisation of plant growth substances based upon the direct coupling of a gas chromatogram, a mass spectrometer and a small computer – recent examples of its application. In *Plant Growth Substances 1970*, ed. D. J. Carr, pp. 790–7. Berlin: Springer.

MacMillan, J. & Wels, C. M. (1973). Partition chromatography of gibberellins and related diterpenes on columns of Sephadex LH-20. *Journal of Chromatography*, **87**, 271–6.

Palmer, L. & Holmstedt, B. (1975). Mass fragmentography – the use of the mass spectrometer as a selective and sensitive detector in gas chromatography. *Science Tools*, **22**, 25–39.

Powell, L. E. & Tautvydas, K. J. (1967). Chromatography of gibberellins on silica gel partition columns. *Nature, London*, **213**, 292–3.

Pryce, R. J. (1973). Decomposition of aqueous solutions of gibberellic acid on autoclaving. *Phytochemistry*, **12**, 507–14.

Railton, I. D. & Wareing, P. F. (1973). Effects of daylength on endogenous gibberellins in leaves of *Solanum andigena*. 1. Changes in levels of free acidic gibberellin-like substances. *Physiologia Plantarum*, **28**, 88–94.

Reeve, D. R. & Crozier, A. (1975). Gibberellin bioassays. In *Gibberellins and Plant Growth*, ed. H. N. Krishnamoorthy, pp. 35–64. New Delhi: Wiley Eastern.

Reeve, D. R. & Crozier, A. (1976). Purification of plant hormone extracts by gel permeation chromatography. *Phytochemistry*, **15**, 791–3.

Reeve, D. R. & Crozier, A. (1977). Radioactivity monitor for high performance liquid chromatography. *Journal of Chromatography*, **137**, 271–82.

Reeve, D. R., Yokota, T., Nash, L. J. & Crozier, A. (1976). The development of a high performance liquid chromatograph with a sensitive onstream radioactivity monitor for the analysis of ^{3}H and ^{14}C-labelled gibberellins. *Journal of Experimental Botany*, **21**, 1243–58.

Rivier, L., Milon, H. & Pilet, P. E. (1977). Gas chromatography–mass spectrometric determinations of abscissic acid levels in the cap and the apex of maize roots. *Planta, Berlin*, **134**, 23–7.

Rivier, L. & Pilet, P. E. (1974). Indolyl-3-acetic acid in cap and apex of maize roots: identification and quantification by mass fragmentography. *Planta, Berlin*, **120**, 107–112.

Snyder, L. R. (1971). The role of the mobile phase in liquid chromatography. In *Modern Practice of Liquid Chromatography*. ed. J. J. Kirkland, p. 125–57. New York: Wiley-Interscience.

Takahashi, N., Murofushi, N., Tamura, S., Wasada, N., Hoshino, H. & Tsuchiya, T. (1969). Mass spectrometric studies of giberellins. *Organic Mass Spectrometry*, **2**, 711–22.

Yamaguchi, I., Yokota, T., Murofushi, N., Ogawa, Y. & Takahashi, N. (1970). Isolation and structure of a new gibberellin from immature seeds of *Prunus persica*. *Agricultural and Biology Chemistry*, **34**, 1439–41.

P. GASKIN & J. MACMILLAN

GC and GC–MS techniques for gibberellins

Summary

GC–MS has the following advantages. First, it provides conclusive identification without the need for reference samples if reference spectra are available. Secondly, it is a sensitive technique. Thirdly, since it involves a separation step it can be applied to mixtures. Fourthly, it can be used for sensitive quantitative analysis, provided suitable standards are available for calibration.

The main disadvantages are the capital cost and the fact that it cannot be applied directly to involatile GA-conjugates.

The main practical problems, however, lie less in GC–MS than in the initial preparation of samples.

Introduction

The use of combined gas chromatography–mass spectrometry (GC–MS) has opened up new vistas in plant hormone research. Although the application of GC–MS analysis to plant hormones was initially developed for gibberellins (GAs) (MacMillan, 1968, 1972), the technique can be applied to other plant hormones, e.g. the cytokinins (Horgan et al., 1973a, b), abscisic acid (Gaskin & MacMillan, 1968), and indole-3-acetic acid (Rivier & Pilet, 1974). GC–MS provides the specificity that is lacking in the simpler techniques of column, thin-layer (TLC) and gas–liquid chromatography (GLC), and of bioassay. Also bioassay is only relevant to biologically active compounds and is limited to relatively pure fractions freed from the growth inhibitors normally present in crude plant extracts.

The main aim of this article is to provide for the untutored beginner a simple guide to the procedures for the analysis of GAs by GC–MS. Attention will be directed to the steps following extraction and partial purification of the extracts. There are, however, some aspects of these earlier steps that need to be mentioned since they will influence the quality of the eventual GC–MS results. Although discussion will be centred on GA analysis, many of the points apply also to the other groups of plant hormones.

Sample preparation
Contaminants

Since GAs are usually present in plant extracts in microgram quantities, contamination from extraneous sources becomes a serious problem. It is vital, therefore, that extracts and samples should be exposed to only glass, Teflon or aluminium foil. Plastic vessels (excluding Teflon) must not be used at any stage. The use of grease for stopcock taps and of the commercial sealing sheet Parafilm should be avoided. Parafilm is composed largely of n-alkanes that are readily soluble in organic solvents (Gaskin, MacMillan, Firn & Pryce, 1971) (Fig. 1). All apparatus must be dedicated to GA analysis and never used for synthetic chemistry, especially of GAs. This latter point is very important. The sensitivity of GC–MS has resulted, in our experience, in the detection, in plant extracts, of some strange compounds which, on further enquiry, were found to arise from communal glassware cleaned in detergent. Solvents must, without exception, be redistilled (Martin, Dennis, Gaskin & MacMillan, 1975) and stored in glass vessels, the caps of which may be Teflon-lined (e.g. Sovirel bottles). Because of their widespread usage, plasticisers are almost

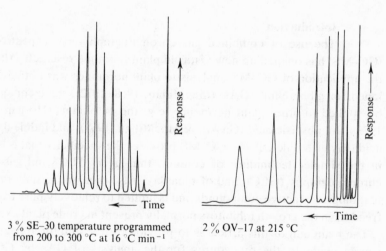

3 % SE–30 temperature programmed   2 % OV–17 at 215 °C
from 200 to 300 °C at 16 °C min⁻¹

Fig. 1. FID GLC traces of n-alkanes from 'Parafilm'.

impossible to exclude completely. For example, polyvinylchloride (PVC) tubing contains nearly 50% by weight of acetyl-tri-*n*-butyl-citrate (Citroflex A-4) (Binks, Goodfellow, MacMillan & Pryce, 1970). This substance chromatographs underivatised and the commercial grade encountered gives rise to one major and several minor peaks on GC. Phthalate esters are notorious for their frequent occurrence in extracts (Hunneman, 1968; Hayashi, Asakawa, Ishida & Matsuura, 1967). Dibutyl and dioctyl phthalates are the most common, although many others are used, and these plasticisers can swamp the most important regions of a GC trace. TLC silica and precoated TLC plates release soluble contaminants and therefore must be pre-eluted with developing solvent before use (Martin *et al.*, 1975). The developing solvent has been found to bring about the release of contaminants from the silica layer. Elution of GAs from TLC plates is usually effected with water-saturated ethyl acetate, methanol or water. The silica should not be separated from the eluting solvent by filtration through filter paper or cotton wool. The method of choice is probably centrifugation.

Many naturally occurring substances that would be expected to separate from GAs do not completely do so on fractionation. As they are present in much larger amounts than the GAs, these substances often complicate an analysis. Fortunately many of them have readily recognisable mass spectra after derivatisation. Examples are (*a*) phenolic acids which are not always removed by polyvinylpyrrolidone (PVP) treatment and which further complicate matters because they often each give rise to two or three derivatives on methylation (Dalgliesh *et al.*, 1966); (*b*) non-polar fatty acids, notably palmitic and stearic acids which occur together with their conjugates. These conjugates appear to be carboxylic acid derivatives as they are not methylatable but yield traces of the trimethylsilyl esters of the free acids after silylation; and (*c*) sugars which GC after silylation and, although they have readily recognisable spectra, also have similar retention times to GAs and mask important features of GA spectra.

To summarise, contaminants which remain undetected in, for example, bioassays can present serious problems in GC and GC–MS analyses. In the case of extraneous contaminants it is useful to run a blank of the complete procedure and if contaminants are detected their source may be recognised (and eliminated) by running blanks of each step. In the case of endogenous interfering substances, experience with a particular type of extract is often the best guide. The more common contaminants show up repeatedly in analyses and become readily recognisable.

Extraction

Common solvents for extraction are methanol, ethanol, ethyl acetate, acetone, *n*-butanol and light petroleum. Care should be exercised when using acetone as a solvent for extraction or chromatography, as it readily forms acetonides with *vic*-diols under mildly acidic conditions (Yamaguchi *et al.*, 1975; Gaskin & MacMillan, 1975). A simplified scheme (Fig. 2) of the extraction procedure in common use gives neutral/basic, acidic, acidic *n*-butanol and water-soluble fractions. As mineral acid is frequently used to adjust the pH of an extract to 2.5 it is necessary to backwash the ethyl acetate with water, otherwise, on evaporation of the organic phase, the resulting concentrated acid will attack any GAs present and cause chemical changes (e.g. rearrangement, hydration, etc.). Drying of organic extracts with drying reagents such as magnesium sulphate is not recommended. The organic extract should be concentrated under reduced pressure and at low temperature. The last traces of water and any remaining acid can be conveniently removed by azeotropic distillation with toluene.

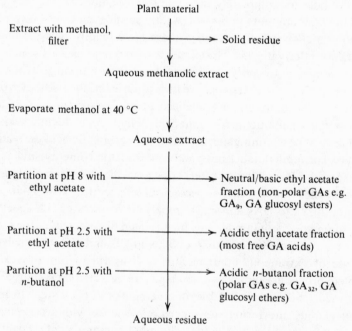

Fig. 2. Simplified scheme for the extraction of gibberellins.

Fractionation

In favourable cases, such as extracts from immature seed, the concentration of GAs in the crude ethyl acetate-soluble fraction may be high enough for GC–MS without further fractionation. However, in most cases the separation of GAs from non-GAs is necessary. There is no ideal solution. Methods depend upon individual situations. In principle, a separation of GAs as a group is all that is required, since the GAs can be separated from each other by GC during GC–MS. However, there is a wide range of polarity within the GA group and a group separation is not readily achieved. Many fractionation procedures have been described and experimentation is usually necessary to determine the best method for particular extracts.

If the amount of extract is small, TLC may be sufficient. Silica gel layers are most commonly used and many solvent systems have been described for development of the layer. The best solvent for elution of GAs from silica gel is ethyl acetate saturated with water. With other solvents such as water and methanol the eluant contains finely divided silica which is difficult to remove and which interferes with derivatisation for GC. The silica particles may be separated by centrifugation. If the total weight of extract is large, column chromatography may be necessary. Most of the published procedures have concentrated on the separation of individual GAs and often only of GA standards. Partition chromatography seems to provide the best separation (see Pitel, Vining & Arsenault, 1971; MacMillan & Wels, 1973) and high performance liquid chromatography (HPLC) provides excellent separation of GAs and a high peak capacity (see Reeve & Crozier, this volume). The advantage of such methods is that fractions containing individual GAs are cleanly separated from non-GAs and from each other. The disadvantage for GC–MS is that many fractions are obtained which have to be individually analysed. Simpler methods such as charcoal-celite columns are often useful for GC–MS.

Conjugates

GA conjugates can be gas chromatographed with difficulty (see later). In crude extracts they may be cleaved by hydrolysis to yield either the free GA or the transformed GA, either of which may then be identified by GC–MS. There are three hydrolytic procedures:

(1) Enzymatic hydrolysis. Hydrolysis using commercial preparations of cellulase is commonly used, although the efficacy of cellulase and β-glucosidase depends upon the source (Knöfel, Müller & Sembdner, 1974). In our hands Boots Pectolytic Enzyme has proved useful for crude extracts. The crude pectinase is washed off the Kieselguhr support with

pH 8 phosphate buffer and incubated at 37 °C for 40 h with a solution of the extract in pH 8 phosphate buffer. Partitioning at pH 2.5 with ethyl acetate yields the free GAs in the organic phase.

(2) Acid hydrolysis. The dried extract in 1 mol dm⁻³ HCl is heated at 100 °C for 3 h. After rapid partitioning with ethyl acetate (to avoid hydrolysis of the ethyl acetate), the acid hydrolysis products in the organic layer are derivatised for GC–MS. The hydrolysis products of the GA conjugates are identified by reference to spectra of the products of acid treatment of GA standards. Bioassay cannot be used here as the treatment usually results in inactivation of any GAs present.

(3) Base hydrolysis. The dried extract in 2 mol dm⁻³ potassium hydroxide is heated at 100 °C for 1 h, the pH taken to 2.5 (with concentrated hydrochloric acid) and the free GAs partitioned into ethyl acetate. Some GAs, e.g. GA_1 and GA_4, rearrange in base to give their 3-epimers.

These three hydrolysis techniques have been successfully used in an analysis of the GA conjugates in a pH 2.5 n-butanol extract of the immature seeds of *Phaseolus coccineus* (Gaskin & MacMillan, 1975).

Derivatisation

GAs are carboxylic acids and decompose under the usual GC conditions. It is necessary therefore to use volatile, thermally stable derivatives for GC.

Methylation

The first step is usually methylation of the carboxylic group to form the methyl (Me) ester as follows: the dried extract is taken up in a minimum of methanol and a freshly made solution of diazomethane in ether is added dropwise until the yellow colour of the reagent persists for 15 to 30 min. Solvent and diazomethane are then removed with gentle warming under a stream of nitrogen gas. All procedures involving diazomethane must be carried out in a fume cupboard as this reagent is explosive and carcinogenic. The best vessel for methylation is a small (½ or 1 dram) screwcap vial. The preparation of diazomethane is described by Vogel (1956) and by Schlenk & Gellerman (1960). Care should be exercised to avoid the carry-over of potassium hydroxide in the distillate, since the presence of this base in the diazomethane solution can catalyse chemical transformations of some GAs (see for example Fig. 3).

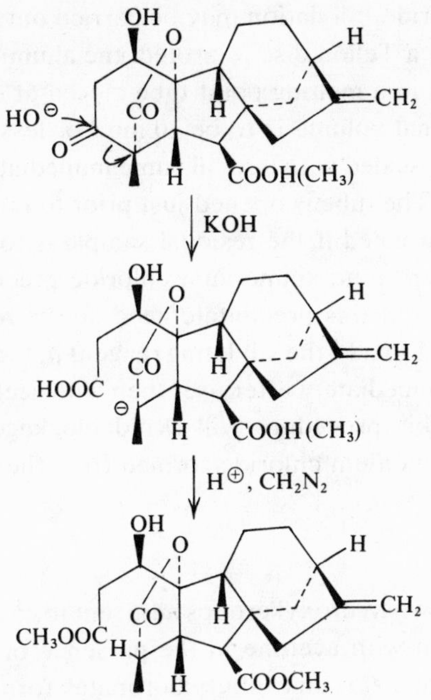

Fig. 3. The reaction of GA_{33} with potassium hydroxide-contaminated ethereal diazomethane.

Silylation

There are a number of silylating reagents but trimethylsilylation has been most commonly used for GAs. Methyl esters of hydroxy GAs are converted to their trimethylsilyl ethers (Me-TMSi) and free GAs to their trimethylsilyl ester trimethylsilyl ethers (TMSi ester TMSi ether). The latter derivatives have the added advantage that they are readily hyrolysed to the free GAs after preparative GC (Lorenzi, Horgan & Heald, 1975). Derivatisation is best carried out on small aliquots of extract because TMSi derivatives, particularly the TMSi esters, have limited stability in the presence of moisture. The extract (methylated or not) is dried *in vacuo* over potassium hydroxide, taken up in dry pyridine and a silylating reagent added. Convenient trimethylsilylating reagents are (*a*) a 3:2:2 (v:v:v) mixture of hexamethyldisilazane, trimethylchlorosilane and pyridine (which is added to the pyridine solution of the extract to give a solvent:reagent ratio of *ca* 1:1 (v:v)); (*b*) BSA (*bis*-trimethylsilylacetamide); and (*c*) BSTFA (*bis*-trimethylsilyltrifluoroacetamide). The reaction mixture is either allowed to stand for 1 h at room temperature or is warmed to *ca* 120 °C for a few minutes. The pyridine must be dried and this is conveniently done by redistillation over potassium hydroxide and storage over calcium hydride. Silylation may be carried out either in a ½ dram screwcap vial (with a Teflon disc to protect the aluminium foil in the cap from reagent *a*) or in a melting point tube closed at one end. The latter is used when the final volume is to be 20 mm³ or less and the open end of the tube should be sealed with a small flame immediately after addition of silylating reagent. The tube is opened just prior to removal of sample for GC and may be resealed if the residual sample is to be kept for further work. Using reagent *a* an ammonium chloride precipitate is produced causing cloudiness, but this precipitate need not be removed. (NB: If a micro-syringe is used to add the silylating reagent *a*, the syringe should be rinsed with water immediately after use, then with acetone and finally, with dry pyridine. This procedure will avoid blockage of the micro-syringe needle with ammonium chloride formed from the reaction of moist air with reagent *a*).

Other derivatives

For gibberellins with two hydroxyl groups adjacent to each other, acetonides formed by reaction with acetone in the presence of an acid catalyst (Gaskin & MacMillan, 1975) and *n*-butylboronates formed with *n*-butylboronic acid (Brooks & Watson, 1967; Brooks & Harvey, 1971) are useful derivatives. For GAs containing keto or aldehyde groups, oximes prepared with hydroxylaminehydrochloride or *O*-methyloximes prepared with methoxyamine hydrochloride (Fales & Luukainen, 1965) are useful derivatives. Oximes may be trimethylsilylated to form the *O*-TMSi-oximes. Trifluoroacetyl derivatives have been used for electron capture detection by GC (Seeley & Powell, 1974).

GC technique

Since GC–MS conditions should be as close as possible to those used for GC, it is important to establish the optimum combination of derivative and GC conditions prior to GC–MS. Columns should be of glass, and for packed use, of 1 m to 3 m in length and of 4 mm or 2 mm internal diameter. Before being packed, the columns should be deactivated by treatment with dimethyldichlorosilane in dry toluene, followed by rinsing with methanol (to react with any residual Si–Cl groups). The liquid phase is usually one of the wide range of silicones, because of their good temperature stability. Most frequently used phases are: SE-30, OV-1, SE-33, OV-17, QF-1, OV-210, OV-225 and XE-60 in increasing order of polarity. They are coated to *ca* 2% (w/w) on a demineralised and deactivated diatomaceous support such as Gas Chrom 'Q' (80–100 or

100–120 mesh). Columns should be packed to *ca* 5 mm above the injection point so that samples can be injected directly onto the packing material. Before use the packed columns must be conditioned in a slow stream of carrier gas (10 cm³ min⁻¹) by temperature programming at 1 or 2 °C min⁻¹ from 100 to 25 °C above their maximum operating temperature and kept at the higher temperature for 24 h. Conditioning is completed by successive injections of several microlitres of *bis*-trimethylsilylacetamide (BSA) at 150 °C followed by temperature programming at 8 °C min⁻¹ to their maximum operating temperature. During conditioning the outlet end of the column should not be connected to the detector. Indeed, when large volumes of silylating agent or solvent are injected, either the column outlet should be disconnected or the flame-ionisation detector should be left unlit.

Examples of GC conditions used for Me-TMSi derivatives of GAs on packed columns are: (1) 2% SE-33 on 80–100 mesh Gas Chrom 'Q' in a 2 m by 2 mm i.d. column with nitrogen carrier gas at a flowrate of 25 cm³ min⁻¹. Following injection the oven temperature is held at 190 °C for 5 min then temperature programmed at 3 °C min⁻¹ to 280 °C; and (2) 2% QF-1 on 80–100 mesh Gas Chrom 'Q' in a 1.5 m by 2 mm i.d. column, nitrogen flow rate 25 cm³ min⁻¹, oven temperature 170 °C for 5 min then programmed at 2 °C min⁻¹ to 250 °C. In both cases injector block and detector oven temperatures are 250 and 280 °C respectively. With 4 mm i.d. columns 40–70 cm³ min⁻¹ nitrogen flow rate is used. The volume injected may vary from 0.1 mm³ to *ca* 8 mm³.

Conditions have been reported for the GC of GA glucosyl ethers as the Me-TMSi and TMS ester TMSi ether derivatives and of GA glucosyl esters on the TMSi derivatives (Schneider, Jänicke & Sembdner, Yamane & Takahashi, 1975; Hiraga, 1974). However, some decomposition occurs under the high column temperatures that are required (Lorenzi *et al.*, 1976). Moreover these derivatives do not readily pass through a molecular separator in GC–MS and the mass spectrum obtained is more characteristic of the carbohydrate portion than of the GA.

Although most GC studies have used packed columns, GA derivatives separate well in wall-coated glass capillary columns.

GC used alone as a means of identification is inconclusive because many other compounds chromatograph with similar retention times to GAs and chromatograms from crude extracts may be very complex. GC retention times do not constitute a conclusive GA identification, although preparative GC followed by bioassay of a trimethylsilylated extract can be helpful in locating a biologically active GA peak in a complex GC trace (Lorenzi *et al.*, 1975).

GC–MS

In GC–MS a fast-scanning magnetic sector or a quadrupole mass spectrometer is directly linked to the gas chromatograph and acts as a sophisticated GC detector. The effluent from the GC column is led into the source of the mass spectrometer either directly or through a device which removes most of the GC carrier gas. Compounds entering the source are ionised and the ions are analysed, as described later, to provide the mass spectrum. In some instruments a portion of the ions is intercepted before analysis to provide a total ion current (TIC) profile (Fig. 4) which is a measure of the amount of compound in the source at any given time and which corresponds to more conventional methods of GC detection such as flame ionisation. In other instruments the TIC trace is reconstructed by summing the total of ions detected after analysis. A further degree of sophistication is the addition of a computer to acquire and

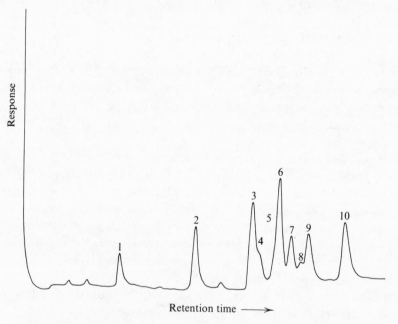

Fig. 4. TIC trace of a mixture of derivatised standards. GC–MS conditions: 2% SE-33 on 80–100 mesh Gas Chrom 'Q' in a 170×0.2 cm i.d. column. Temperature programmed from 190 to 280 °C at 6 °C min^{-1}. Helium flow rate 25 cm^3. Separation temperature 190 °C. Source temperature 210 °C. Electron energy 24 eV.

1 = Me-ABA, 2 = GA$_9$Me, 3 = GA$_4$+GA$_7$ Me-TMSi,
4 = GA$_7$Me-TMSi, 5 = GA$_{17}$Me-TMSi, 6 = GA$_{13}$Me-TMSi,
7 = GA$_3$–19,2 lactone Me-TMSi, 8 = GA$_1$Me-TMSi, 9 = GA$_3$Me-TMSi,
10 = GA$_8$Me-TMSi.

process the vast amount of data that a GC–MS system can produce.

By far the most common method for the production of ions in the mass spectrometer is electron bombardment of the molecules in the gaseous phase, usually called electron-impact (EI). Other methods such as the less energetic chemical ionisation (CI) and field ionisation (FI) have not yet been used in GC–MS of GAs. In EI the energy imparted to the neutral molecules is sufficient to cause extensive fragmentation to give many ions. These ions are analysed according to their mass to charge ratio (m/e value) and their relative abundance, and these data are recorded to give the mass spectrum. The latter, together with GC retention time, provide a unique identification of individual compounds.

Many integrated GC–MS systems are now available commercially. There are many options enabling either packed or capillary columns to be used with or without molecular separators. The sensitivity of detection depends upon many factors (type of compound, GC-resolution, mode of operation) but primarily upon the mass spectrometer. The limits of detection are discussed later. The examples used to illustrate the way GC–MS can be applied to GAs are taken from the authors' own experience using packed columns in a Pye 104 gas chromatograph coupled through a single stage silicone membrane separator to an AEI MS30 mass spectrometer. The mass spectral data are acquired and processes by a DEC Linc 8 computer using 'home-brewed' programs.

Having previously determined the optimum GC conditions for the derivatised GC sample the MS operation will depend upon the complexity of the sample and the nature of the information required. There are three options.

Manual scanning

A single scan may be initiated manually when the TIC trace reaches a maximum of intensity. The spectrum can be recorded through a galvanometric recorder on UV-sensitive paper (Fig. 5) from which the m/e values are counted and their intensities measured by hand. Alternatively the data may be collected and processed directly by a computer to give the mass spectrum in the form of a line diagram (Fig. 6) where the intensities of the ions are expressed as percentages either of the most intense peak (base peak) or of the total ion current. Manual scanning is generally used for standard compounds, synthetic samples or simple mixtures. To obtain an identifiable mass spectrum, the minimum amount varies between 5×10^{-8} and 5×10^{-11} g, depending upon the GC–MS system used, for each component in the injected sample.

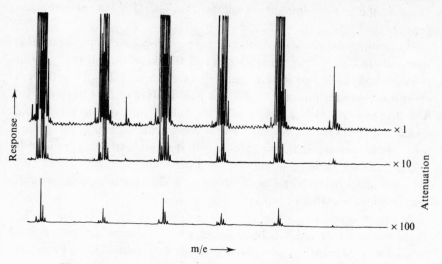

Fig. 5. A portion of a mass spectrum recorded by a galvanometric recorder on UV-sensitive paper.

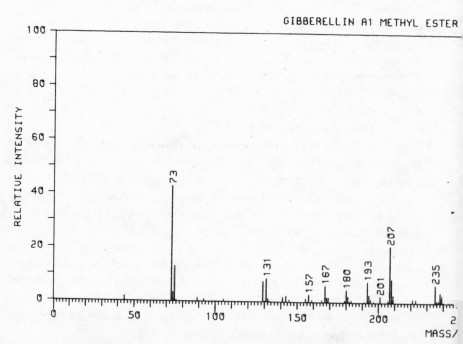

Fig. 6. Gibberellin A₁ methyl ester TMS ether. MW = 506.

Repetitive scanning

In this method the mass spectrometer automatically and continuously scans throughout the GC run. This mode of operation requires a computer which is used to initiate repetitive scanning and to process the mass spectral data. For crude plant extracts this is the most useful method as the maximum amount of information is recorded for further analysis by the computer in many sophisticated ways. For example, the computer can generate mass fragmentograms which show the distribution and intensities of ions of selected m/e values throughout the GC–MS run. Minor components of a complex mixture can be detected in this way and then identified. A good example is provided by the identification of GA_{29} Me-TMSi in a methylated and trimethylsilylated extract of the flesh of prune fruit (Fig. 7). The ions at m/e 506 (M^+) and m/e 416 (M^+-TMSiOH), present in the reference spectrum of GA_{29} Me-TMSi, can be seen to reach a maximum at scan No. 119 in the mass fragmentogram (Fig. 7) and the full mass spectrum of scan No. 119 from which the mass spectrum from scan No. 116 had been subtracted was identifiable as GA_{29} Me-TMSi. The ions at m/e 506 and 416, present in scan No. 112, belong to a dihydroxy GA_9 which has not yet been identified. Given good GC resolution and the absence of interfering ions 50 ng of a GA can be recognised in an injection of several milligrams of a plant extract by this method.

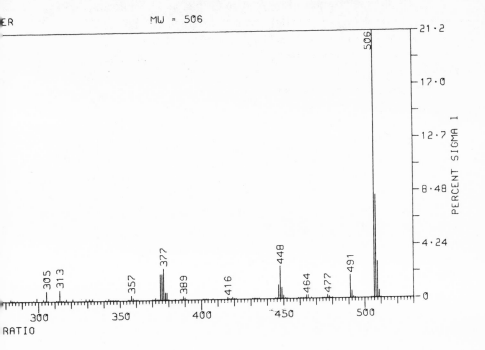

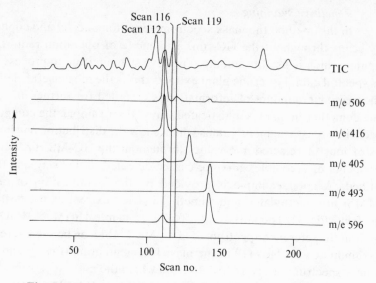

Fig. 7. Reconstituted total ion (TIC) and single ion (mass fragmentogram) traces from repetitive scan GC–MS of a derivatised (Me-TMSi) extract of the flesh of prune fruit.

Selected ion current monitoring (SICM)

In this method (also known as multiple peak monitoring (MPM), multiple ion detection (MID), mass fragmentography (MF), and simultaneous-ion monitoring (SIM)) the intensity of one or more selected ions is directly recorded during a GC–MS run. The mass spectrometer is first tuned to focus the selected m/e values using standard compounds which produce these ions. This method can be used to identify known GAs if a sufficient number (usually six) of the most characteristic ions are recorded. However, the method is generally used for quantitative analysis using one selected ion. For quantitative analyses suitable standards are required. The simplest method uses only an external standard which should preferably be the GA being analysed. The response of the GA in the extract is related to that of the external standard to give an estimate of the GA present in the extract. A more accurate method is to add an internal standard which gives an ion at the same m/e as that being monitored for the GA and which has a different GC retention time. This corrects for variation in injection size and selected ion current response. The same amount of this internal standard is added to a range of concentrations of the GA external standard and selected ion current responses for these mixtures provide a calibration for quantification.

If the GA and internal standard have similar physical properties, the

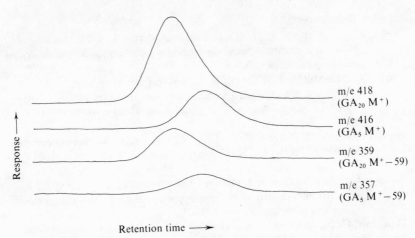

Fig. 8. SICM trace of a mixture of GA_{20} Me-TMSi and GA_5 Me-TMSi (QF-1, 190 °C).

internal standard can be added at the beginning of the extraction and the percentage recovery of the standard, and hence of the GA, can be determined. The limits of detection by SICM depend upon the sensitivity of the mass spectrometer and the method of ionisation. Using EI and the most abundant (intense) ion, 5×10^{-11} to 5×10^{-14} g can be detected.

SICM can also be used to resolve components of a mixture that are incompletely resolved by GC (Fig. 8) and to quantify each of them separately.

In order to identify known GAs, the mass spectrum obtained by GC–MS is compared with reference spectra. Reference spectra for the Me and Me-TMSi derivatives of GA_1 to GA_{24} have been published (Binks *et al.*, 1969). Reference spectra for the Me, Me-TMSi, and TMSi ester TMSi ether derivatives of all the known GAs are in the process of being published. For quantitative work, however, samples of authentic GAs and of suitable internal standards are necessary.

In addition to providing the identification of known GAs, GC–MS can also detect new GAs to which structures may often be assigned from their mass spectra. However, the correlation of fragmentation patterns and GA structures is beyond the scope of this article.

References

Binks, R., Goodfellow, R. J., MacMillan, J. & Pryce, R. J. (1970). Acetyl tri-*n*-butyl citrate, a common laboratory contaminant. *Chemistry and Industry (London)*, 565–6.

Binks, R., MacMillan, J. & Pryce, R. J. (1969). Plant hormones VIII. Combined gas chromatography–mass spectrometry of the methyl esters of gibberellins A_1 to A_{24} and their trimethylsilyl ethers. *Phytochemistry*, **8**, 271–84.

Brooks, C. J. W. & Harvey, D. J. (1971). Comparative gas chromatographic studies of corticosteroid boronates. *Journal of Chromatography*, **54**, 193–204.

Brooks, C. J. W. & Watson, J. (1967). Characterization of 1,2- and 1,3-diols by gas chromatography–mass spectrometry of cyclic boronate esters. *Chemical Communications*, 952–3.

Dalgliesh, C. E., Horning, E. C., Horning, M. G., Knox, K. L. & Yarger, K. (1966). A gas–liquid chromatographic procedure for separating a wide range of metabolites occurring in urine or tissue extracts. *Biochemical Journal*, **101**, 792–810.

Fales, H. M. & Luukainen, T. (1965). *O*-methyloximes as carbonyl derivatives in gas chromatography, mass spectrometry and nuclear magnetic resonance. *Analytical Chemistry*, **37**, 955–7.

Gaskin, P. & MacMillan, J. (1968). Plant hormones VII. Identification and estimation of abscisic acid in a crude plant extract by combined gas chromatography–mass spectrometry. *Phytochemistry*, **7**, 1699–1701.

Gaskin, P. & MacMillan, J. (1975). Polyoxygenated *ent*-kaurenes and water-soluble conjugates in seeds of *Phaseolus coccineus*. *Phytochemistry*, **14**, 1575–8.

Gaskin, P., MacMillan, J., Firn, R. D. & Pryce, R. J. (1971). 'Parafilm': a convenient source of *n*-alkane standards for the determination of gas chromatographic retention indices. *Phytochemistry*, **10**, 1155–7.

Hayashi, S., Asakawa, Y., Ishida, T. & Matsuura, T. (1969). Phthalate esters of *Cryptotaenia canadensis* DC. var. *japaonica* Makino (Umbelliferae). *Tetrahedron Letters*, 5061–3.

Hiraga, K., Yamane, H. & Takahashi, N. (1974). Biological activity of some synthetic gibberellin glucosyl esters. *Phytochemistry*, **13**, 2371–6.

Horgan, R., Hewett, E. W., Purse, J. G., Horgan, J. M. & Wareing, P. F. (1973*a*). Identification of a cytokinin in sycamore sap by gas chromatography–mass spectrometry. *Plant Science Letters*, **1**, 321–4.

Horgan, R., Hewett, E. W., Purse, J. G. & Wareing, P. F. (1973*b*). A new cytokinin from *Populus robusta*. *Tetrahedron Letters*, 2827–8.

Hunneman, D. H. (1968). A cautionary note on phthalate esters of *Cryptotaenia canadensis*. *Tetrahedron Letters*, 1743.

Knöfel, H.-D., Müller, P. & Sembdner, G. (1974). Studies on the enzymatic hydrolysis of gibberellin-*O*-glucosides. In *Biochemistry and Chemistry of Plant Growth Regulators*, International Symposium Cottbus, GDR, September 1974, ed. K. Schreiber, H. R. Schütte & G. Sembdner, pp. 121–4. Halle (Saale), GDR: Institute of Plant Biochemistry.

Lorenzi, R., Horgan, R. & Heald, J. K. (1975). Gibberellins from *Picea sitchensis* Carriere: seasonal variation and partial characterisation. *Planta, Berlin,* **126,** 75–82.

Lorenzi, R., Horgan, R. & Heald, J. K. (1976). Gibberellin A_9 glucosyl ester in needles of *Picea sitchensis. Phytochemistry,* **15,** 789–91.

MacMillan, J. (1968). Direct identification of gibberellins in plant extracts by gas chromatography–mass spectrometry. In *Biochemistry and Physiology of Plant Growth Substances,* ed. F. Wightman & G. Setterfield, pp. 101–7. Ottawa: Runge Press.

MacMillan, J. (1972). A system for the characterisation of plant growth substances based upon the direct coupling of a gas chromatogram, a mass spectrometer, and a small computer – recent examples of its application. In *Plant Growth Substances 1970,* ed. D. J. Carr, pp. 790–7. Berlin, Heidelberg, New York: Springer.

MacMillan, J. & Wels, C. M. (1973). Partition chromatography of gibberellins and related diterpenes on columns of Sephadex LH-20. *Journal of Chromatography,* **87,** 271–6.

Martin, G. C., Dennis, F. G., Gaskin, P. & MacMillan, J. (1975). Contaminants present in materials commonly used to purify plant extracts for hormone analysis. *Horticultural Science,* **10,** 598–9.

Pitel, D. W., Vining, L. C. & Arsenault, G. P. (1971). Improved methods for preparing pure gibberellins from cultures of *Gibberella fujikuroi.* Isolation by adsorption or partition chromatography on silicic acid and by partition chromatography on Sephadex columns. *Canadian Journal of Biochemistry,* **49,** 185–93.

Rivier, L. & Pilet, P.-E. (1974). Indolyl-3-acetic acid in cap and apex of maize roots: identification and quantification by mass fragmentography. *Planta, Berlin,* **120,** 107–12.

Schlenk, H. & Gellerman, J. L. (1960). Esterification of fatty acids with diazomethane on a small scale. *Analytical Chemistry,* **32,** 1412–14.

Schneider, G., Jänicke, S. & Sembdner, G. (1975). Gibberelline XXXIV. Beitrag zur Gaschromatographie von Gibberellinen und Gibberellin-*O*-Glucosiden-*N,O*-bis(trimethylsilyl)acetamid als Silylierungsreagens. *Journal of Chromatography,* **109,** 409–12.

Seeley, S. D. & Powell, L. E. (1974). Gas chromatography and detection of microquantities of gibberellins and indoleacetic acid as their fluorinated derivatives. *Analytical Biochemistry,* **58,** 39–46.

Vogel, A. I. (1956). *A Text-book of Practical Organic Chemistry,* 3rd edn, pp. 967–73. London: Longmans.

Yamaguchi, I., Yokota, T., Murofushi, N., Takahashi, N. & Ogawa, Y. (1975). Isolation of gibberellins A_5, A_{32}, A_{32} acetonide and (+)-abscisic acid from *Prunus persica. Agricultural and Biological Chemistry, Japan,* **39,** 2399–2403.

R. HORGAN

Analytical procedures for cytokinins

Introduction

Since the cytokinins first emerged as a class of endogenous plant growth substances some twelve years ago, numerous analytical techniques have been described in the literature for their detection, isolation and identification. Space does not permit this review to contain a critical and comprehensive account of the many and varied methods that have been used for cytokinin analysis; instead it is intended to concentrate mainly on those aspects of the subject which in the author's view have proved to be most problematical.

No account will be given of the various bioassay systems used to detect and measure cytokinins in plant extracts as it is felt that this is a suitable subject for a separate review. The emphasis of this article will be on the chemical and biochemical problems associated with the isolation and purification of cytokinins prior to the determination of their structure by spectroscopic methods.

Few of the techniques used for the isolation and identification of cytokinins have been developed specifically for that purpose. In general they have been borrowed from the fields of purine chemistry and bio-chemistry and so the author does not feel that it is necessary to provide comprehensive lists, say of solvents suitable for paper chromatography of cytokinins, since these may be found by reference to standard texts.

Most of the techniques described and assessed in this review have been experienced first hand by the author and his colleagues and it is felt that they can be recommended to other workers in the field because of their reproducibility. However, it should be remembered that as in any field of analytical endeavour where one is attempting to isolate and identify minute quantities of substances from much larger quantities of impurities, attention to the minutest detail is of critical importance and very often 'it ain't what you do but the way that you do it, that's what gets results'.

Extraction of cytokinins from plant material

The traditional method of extracting plant growth substances by homogenisation of the plant tissue in 80% methanol or ethanol at *ca* 4 °C has severe drawbacks when applied to cytokinins.

It has been known for some time that plant tissues contain certain non-specific phosphatases which are not completely inactivated during extraction by the above method. As a result of this the yields of cytokinin nucleotides extracted by this method will be low and possibly very variable. Hence any attempts to read significance into the cytokinin nucleoside:nucleotide ratio will be quite pointless. However, as the outcome of a comprehensive study of plant phosphatases and the problems associated with inactivating them, Bieleski (1964) has developed a solvent 'cocktail' and a method of extraction which is certainly the best available technique for extracting cytokinin nucleotides intact. A major problem associated with the Bieleski extraction method is the large amount of lipid

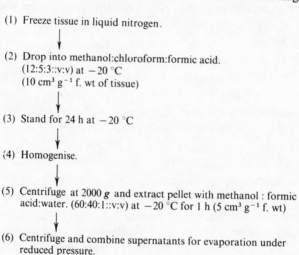

(1) Freeze tissue in liquid nitrogen.

(2) Drop into methanol:chloroform:formic acid.
     (12:5:3::v:v) at −20 °C
     (10 cm³ g⁻¹ f. wt of tissue)

(3) Stand for 24 h at −20 °C

(4) Homogenise.

(5) Centrifuge at 2000 *g* and extract pellet with methanol : formic acid:water. (60:40:1::v:v) at −20 °C for 1 h (5 cm³ g⁻¹ f. wt)

(6) Centrifuge and combine supernatants for evaporation under reduced pressure.

If extraction of a large quantity of lipid material represents a serious problem, after (4) add chloroform and water until ratio of methanol:chloroform:formic acid is 12:11:10. Separate phases by centrifugation and evaporate aqueous phase under reduced pressure.

Fig. 1.

Table 1. *Recovery of [^{14}C]AMP from extracts of soybean callus tissue*

Extraction method	AMP	Adenosine
'Traditional' 80% CH$_3$OH extraction. Tissue residue extracted overnight at 4 °C.	8	92
CH$_3$OH added to make 80% with tissue water. No homogenisation. Stand overnight at 0 °C.	90	10
Bieleski (1964)	88	12
Perchloric acid	91	9

material extracted from the tissue. Where this is a serious problem because of interference with subsequent purification steps, a slightly modified technique may be used. Details of these extraction techniques are given in Fig. 1.

Ice-cold perchloric acid (0.1–1 mol dm^{-3}) has frequently been used to extract nucleotides and Laloue, Terrine & Gawer (1974) has used it as an extraction medium for cytokinin metabolites. In the author's experience some loss of zeatin occurs when ice-cold perchloric acid is used to extract zeatin metabolites. In addition, large scale extractions of plant tissue (> 5 kg fresh weight (f. wt)) are very difficult to work up when perchloric acid

Table 2. *Recovery of metabolites of [^{14}C]zeatin and [^{14}C]benzyladenine from extracts of soybean callus tissue*

	Zeatin			Benzyladenine		
Extraction method	Base	Riboside	Ribotide	Base	Riboside	Ribotide
Traditional 80% CH$_3$OH extraction. Tissue residue extracted overnight at 4 °C.	30	65	5	20	80	0
CH$_3$OH added to make 80% with tissue water. No homogenisation. Stand overnight at 0 °C.	32	0	68	23	10	67
Bieleski (1964)	40	0	60	25	6	69
Perchloric acid	8[a]	0	78	23	3	74

[a] This figure is the result of degradation of zeatin to several unidentified compounds, which represent *ca.* 10% of the total radioactivity.

is used as the extraction medium. This is a consequence of the large amount of potassium perchlorate precipitated during the neutralisation step.

A simple extraction technique suitable for soft tissues (e.g. callus tissue) is to add the tissue to ethanol at 0 °C so as to make the solution 80% in ethanol with the tissue water and allow to stand at 0 °C for 12 h. In the author's experience, provided that the tissue is not homogenised, very little breakdown of nucleotides occurs with this technique and overall recovery of cytokinins is > 80%. Tables 1 and 2 illustrate the effects of different extraction techniques on the recovery of nucleotides.

In conclusion, the author feels he should stress that proper attention must be given to the choice of extraction medium if reliable results regarding the amounts of cytokinin nucleosides/nucleotides present in plant tissues are to be obtained.

Ion exchange methods

The use of an ion exchange step is usually obligatory as a preliminary clean-up during the isolation of cytokinins from plant tissue. The most widely used ion exchange materials have been Dowex 50 and Dowex 1 or their equivalents, and cellulose phosphate and DEAE cellulose.

The methods of use of these ion exchange materials are comprehensively covered in the manufacturers' literature and need not be discussed here. However, some specific points regarding their use for cytokinin isolation need to be considered.

It is clear that the recoveries of small quantities of cytokinin bases and nucleosides from strong cation exchange resins are not quantitative and this problem has been investigated by Vreman & Corse (1975). In the author's experience the recovery of zeatin and its metabolites from Zerolit 225 cation exchange resins is somewhat variable and in the range 50 to 80% if 1 mol dm^{-3} NH$_4$OH is used as the eluting solvent. The recovery depends on the initial purity of the sample and the batch of ion exchange resin used. However, the use of 50% methanol in 1 mol dm^{-3} NH$_4$OH as eluting solvent leads to consistent recoveries of > 90%.

When working with strong cation exchange resins of the Dowex 50 type, considerable care should be exercised over the choice of ionic form used. Several workers have used these resins in the H$^+$ form in spite of the evidence of cytokinin breakdown under these conditions (Dyson & Hall, 1972). These results are best used in the NH$_4^+$ form for cytokinin work. However, strong cation exchange resins and celluloses in the NH$_4^+$ form are subject to hydrolysis with the release of ammonium hydroxide. Thus the pH of columns prepared from strong cation exchanges in the NH$_4^+$

form can rise if they are allowed to stand for any length of time. This may result in loss of ion exchange capacity if the pH of the column is not adjusted before application of the sample. In addition, since the pK_a of most cytokinins is approximately 4 in water, the pH of samples applied to cation exchange is *ca* 3. However, when aqueous methanolic extracts are applied directly to cation exchange columns it should be realised that the pK_as of cytokinins under these conditions will be < 4.

Vreman & Corse (1975) have advocated use of the weak cation exchange resin Duolite CS-101 with cytokinin samples applied in aqueous solution at pH 10. Excellent recoveries of bases and nucleosides were reported when columns were eluted with four bed volumes of 1 mol dm^{-3} NH$_4$OH in 70% ethanol. Since cytokinin nucleosides will be uncharged and bases will be partially in the anionic form at pH 10, it is not clear by what mechanism this column is operating.

In the author's experience cellulose phosphate is the most suitable cation exchange material for cytokinin applications. Excellent recoveries > 90% of zeatin and related compounds are obtained from columns of cellulose phosphate (NH$_4$$^+$) washed with five bed volumes of water at pH 3 (acetic acid) and eluted with five bed volumes of 1 mol dm^{-3} NH$_4$OH.

The recovery of cytokinin nucleotides from anion exchange columns has not been studied to any great extent. The author has found that zeatin ribotide is quantitatively recovered from columns of Dowex 1 (HCOO$^-$ form) eluted with 1 mol dm^{-3} HCOOH and from columns of DEAE cellulose (HCO$_3$$^-$ form) eluted with 1 mol dm^{-3} NH$_4$HCO$_3$.

It should be noted that when using both anion and cation exchange columns to fractionate the cytokinins into nucleotides and bases/nucleosides it is advisable to carry out the cation exchange step first. At the pH used to apply the material to the anion exchange column, usually *ca* pH 8 for DEAE cellulose in the HCO$_3$$^-$ form, there will be sufficient of the bases in the anionic form to retard them substantially on the column. Hence they may not be completely removed from the column by the initial washing and so may elute with the nucleotide fraction.

In the author's opinion the use of cellulose ion exchange materials is to be preferred over the use of resins. This is principally because other incompletely understood factors besides ion exchange are involved in the binding of cytokinins to ion exchange resins. Details of the use of cellulose phosphate and DEAE cellulose for the isolation of cytokinins may be found in several papers by Letham and his co-workers (Parker & Letham, 1973a; Letham, 1973).

Solvent partition

Some form of solvent partitioning is usually essential as a preliminary clean-up during cytokinin isolation. The use of organic solvents to extract cytokinins and/or remove impurities from aqueous extracts containing cytokinins has been subject to considerable confusion in the literature. This is particularly true of the use of ethyl acetate to extract selectively impurities from aqueous extracts at pH 2.5.

Before deciding to use a particular solvent partitioning method, workers are urged to consult the published work on the subject. Letham (1974) has given reliable data on the partition coefficients of a limited number of cytokinins between several organic solvents and water at different pHs. Table 3 shows data obtained by Purse (1974). In spite of the inherent inaccuracies of the method used (UV absorbance), these data have proved to be very reliable as a guide to the right organic solvent and pH to use to extract various cytokinins; they are of great assistance when used in conjunction with LH-20 Sephadex chromatography (see next section).

It can be seen clearly by reference to Table 3 that clean separation of cytokinins is unlikely to be achieved by solvent partitioning and the method should only be considered as a clean-up procedure.

Chromatography of cytokinins
Column chromatography

The most widely used material for column chromatography of cytokinins has been LH-20 Sephadex. The development of LH-20 Sephadex columns using 35% ethanol as the eluant (Armstrong *et al.*, 1969) produced an extremely useful method for the purification of cytokinins. Although originally developed for the isolation of cytokinin nucleosides from t-RNA hydrolysates the technique is equally applicable to plant extracts. Several authors have reported on the use of this technique to isolate free cytokinins from plant tissue (Laloue, Gawer & Terrine, 1975; Horgan *et al.*, 1975; Wang, Thompson & Horgan, 1977).

The technique was originally described as adsorption chromatography, however, by reference to the elution volumes of some common cytokinins given in Table 4 and the partition coefficient data given in Table 3 it can be seen that the more polar cytokinins elute first from these columns. Thus the mechanism of separation must be reversed phase partition chromatography. The approximate correlation between the position of elution of a cytokinin from a LH-20 Sephadex column with 35% ethanol and its partition coefficient can be used to select the most suitable organic solvent and pH to extract an unknown cytokinin from aqueous solution. For

Table 3. *Mean values of partition coefficients* ($K = C_{org}/C_{aq.}$) *of nine cytokinins between* H_2O *pH 2.5 and 8.2 and petroleum ether, diethyl ether, ethyl acetate and n-butanol*

Organic phase pH	Et₂O[a] 2.5	Et₂O[a] 8.2	EtOAc[a] 2.5	EtOAc 8.2	n-BuOH[b] 2.5	n-BuOH 8.2	Petroleum (60–80) 2.5	Petroleum (60–80) 8.2
Cytokinin								
ms2iP[c]	> 20	> 20	> 20	> 20	> 20	> 20	—	—
ms2iPA[c]	> 20	> 20	> 20	> 20	> 20	> 20	> 0.05	> 0.05
2iP[c]	0.07	2.7	0.3	7.1	5.7	> 20	—	—
2iPA[c]	> 0.05	0.3	0.2	2.6	5.7	> 20	> 0.05	> 0.05
Zeatin	> 0.05	> 0.05	> 0.05	0.2	1.0	11	—	—
Dihydrozeatin	> 0.05	> 0.05	> 0.05	0.1	0.6	8.6	—	—
Zeatin riboside	> 0.05	> 0.05	> 0.05	0.1	1.1	2.1	—	—
Zeatin 9-glucoside	> 0.05	> 0.05	> 0.05	0.09	0.7	0.9	—	—
N⁶-(2-hydroxybenzyl) adenosine	> 0.05	0.1	1.1	4.4	6.7	14	—	—
Kinetin	0.237	0.810	1.78	3.29	8.5	20.6	> 0.05	> 0.05
Kinetin riboside	—	—	6.2	—	—	—	—	—

[a] $Et = C_2H_5-$; $Ac = CH_3COO-$.
[b] $Bu = C_4H_9-$.
[c] See notes to Table 4.

Table 4. *The elution values of some cytokinins and adenine metabolites on Sephadex LH-20*

Compound	Solvent A^a	Solvent B^b	Solvent C^c
Z	—	1.42–1.61	0.89–0.95
ZR	1.87–2.13	1.14–1.30	0.98–1.04
Z9G	1.28–1.45	0.96–1.04	0.94–1.00
iP	6.21–7.06	2.22–2.43	0.83–0.89
iPA	—	1.80–1.92	—
msiPA	—	3.53–3.76	—
Ad	—	≡ Z	1.07–1.16
Ado	—	≡ ZR	1.25–1.35
Gu	—	1.30–1.39	—
Guo	—	0.98–1.14	1.32–1.41
Ino	—	0.65–0.73	1.16–1.22
Z-O-G	1.50–1.80	0.96–1.04	0.94–1.00

KEY: Values are in column volumes (CV) based on the following data.
[a] 2.5×72 cm column; 30 cm^3 h^{-1} flow rate; each fraction = 0.085 CV.
[b] $2.5 \times 75/85$ cm column; 30 cm^3 h^{-1} flow rate; each fraction = 0.0815 and 0.719 CV respectively.
[c] 2.5×100 cm column; 15 cm^3 h^{-1} flow rate; each fraction = 0.036 CV.
Zeatin (Z), zeatin riboside (ZR), zeatin-9-β-D-glucoside (Z9G), N^6-(Δ^2-*iso*-pentenyl)adenine (iP), N^6-(Δ^2-*iso*-pentenyl)adenosine (iPA), 2-methylthio-iPA (msiPA), adenine (Ad), adenosine (Ado), guanine (Gu), guanosine (Guo), inosine (Ino), zeatin-O-glucoside (Z-O-G).
Solvent A = water pH 6.0; solvent B = 35% ethanol; solvent C = 92% ethanol (10^{-3} mol dm^{-3} HCl).

example, if, on a column of the size indicated in Table 4, biological activity was located at 3.0–3.5 column volumes, then ethyl acetate at pH 8.2 could be used as a partitioning solvent in preference to butan-1-ol, with a resulting increase in the initial degree of purification of the material.

Two other solvents used in conjunction with LH-20 Sephadex deserve special mention. First, Burrows *et al.* (1970) used water as eluant to separate N^{-6}-(Δ^2-*iso*-pentenyl)-adenosine (iPA) and 2-methylthio-zeatin riboside (2msZR), since these compounds co-chromatographed when 35% ethanol was used as the eluant. The author and his co-workers have found the LH-20 Sephadex/water combination particularly useful in separating polar cytokinins which elute in the zeatin-9-β-D-glucoside (Z9G) to zeatin riboside (ZR) region on LH-20 Sephadex/35% ethanol columns (see Table 4). This region is found to contain a large proportion of the dry weight of most plant extracts and re-chromatography of any biological activity found in this region on a LH-20 Sephadex/water column can lead to a considerable degree of purification. Wang *et al.* (1977) have successfully

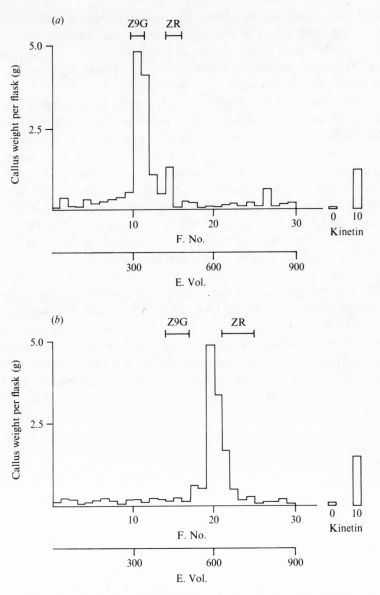

Fig. 2. Chromatography of dihydrozeatin-*O*-β-D-glucoside on columns of LH-20 Sephadex eluted with (*a*) 35% ethanol and (*b*) water. Z9G = zeatin-9-β-D-glucoside; ZR = zeatin riboside. F. No. = fraction number; E. Vol. = elution volume.

used this technique to isolate the new cytokinin dihydrozeatin-O-D-glucoside from leaves of *Phaseolus vulgaris* (Fig. 2).

Secondly, when 92% ethanol (10^{-3} mol dm^{-3}) is used as an eluant for LH-20 Sephadex columns a different and largely unknown mechanism of separation operates (van Es, 1971, personal communication). Under these conditions most of the cytokinins tend to elute close together (see Table 4). In the author's experience this may sometimes result in a very large decrease in dry weight of the active components of the extract. An additional very useful application of such a column is in the study of cytokinin biosynthesis and metabolism. As can be seen from Table 4, this column will completely separate adenine, adenosine, zeatin and zeatin riboside. This is of great value where adenine is used as a labelled precursor for studies into cytokinin biosynthesis, as it enables the bulk of unmetabolised adenine to be quickly and cleanly removed from any labelled zeatin or similar products that may be present in very small quantities.

In practice, column chromatography using LH-20 Sephadex and the solvents previously mentioned is easy to carry out. However, because of the ease of compression of beds of LH-20 Sephadex, columns run by gravity flow invariably compress after a period of time and have to be repacked. A simple, pressurised system incorporating a syringe injector which can be used for as many as 100 consecutive samples is shown in Fig. 3. This system has an added advantage in that the initial compression of the bed under 140 kPa pressure results in increased resolution of cytokinins.

Polyvinylpyrrolidone (PVP) has been used for column chromatography of cytokinins (Glenn, Kuo, Durley & Pharis, 1972; Biddington & Thomas, 1973). A serious disadvantage of this method would seem to be the necessity of using involatile buffer salts as eluants.

Paper and thin-layer chromatography

Both paper chromatography (PC) and thin-layer chromatography (TLC) have been used extensively for analysis of cytokinins. Details of the more commonly used solvent systems will not be given here. A comprehensive list will be found in a paper by Parker & Letham (1973a).

In the author's experience recovery of microgram quantities of cytokinins from Whatman 3MM paper is > 80% if continuous elution with 80% ethanol is used; although for the more polar cytokinins (e.g. glucosides) 50% ethanol is to be preferred and care should be taken not to let the paper dry out. If PC is used as a final purification step prior to spectroscopic analysis it should be realised that even exhaustively washed paper will contribute a background. In the case of UV spectra the reference beam

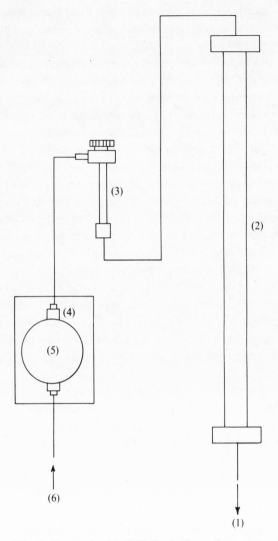

Fig. 3. Pressurised LH-20 Sephadex column set-up. (1) Effluent to UV
monitor and fraction collector; (2) column (Pharmacia SR25/100); (3)
septum injection tee (constructed from Pye 104 GLC injection
head and Chromatronix glass-to-teflon tubing connector); (4) pump to
tubing connector (Chromatronix 1/8 inch National Pipe Thread pipe
connection); (5) solvent pump (Metring pumps series 2 micropump),
solvent flow rate 30 cm³ h⁻¹; (6) from solvent reservoir.

All connections made with Chromatronix tube fittings. Injections
made using 2.5 cm³ Hamilton gastight syringe.

should contain a blank prepared by elution of a piece of blank paper equal in size to the sample zone. This blank can also be used to obtain a background mass spectrum.

TLC has been shown to possess excellent resolving power for cytokinins of very similar structure. For example, the isomeric 3-, 7- and 9-β-D-glucosides of benzyladenine may be separated by TLC as may zeatin-7- and 9-β-D-glucosides (Letham *et al.*, 1975; Parker & Letham, 1973*b*). The *cis* and *trans* isomers of zeatin and zeatin riboside may be completely separated on silica gel thin layers using chloroform:methanol (9:1) as developing solvent (Playtis & Leonard, 1971).

In the author's experience the excellent analytical qualities of TLC are outweighed in its application to preparative problems by the very poor and variable recoveries of microgram quantities of cytokinins from silica gel layers. Whilst microgram quantities of zeatin may be reproducibly recovered from silica gel layers by elution with 80% ethanol (0.01 mol dm^{-3} in acetic acid) with yields of *ca* 80%, recovery of the more polar cytokinins is both low and variable (5–30%). In certain instances unknown polar cytokinins presumed to be glycosides have been lost completely when chromatographed on silica gel layers.

Parker & Letham (1973*a*, *b*) used TLC extensively in their isolation of zeatin metabolites but they did not give details of recoveries.

Gas–liquid chromatography (GLC)

GLC has been used extensively as a technique for the analysis of cytokinins as their trimethylsilyl (TMSi) derivatives. Several silicone-type stationary phases have been used and the separations obtained have been comparable. The author has found 2% OV-1 and 100–200 mesh Gas Chrom Q to be an excellent phase/support combination for GLC analysis of TMSi cytokinins. Columns of this material ($1.5 \times 4 \times 10^{-3}$ m i.d.) will easily resolve *cis* and *trans* zeatin and *cis* and *trans* zeatin riboside as their TMSi derivatives and also separate TMSi *trans* zeatin from TMSi dihydrozeatin and TMSi *trans* zeatin riboside from TMSi dihydrozeatin riboside. The latter two pairs of compounds are impossible to separate by other means. In addition OV-1 has a high temperature stability and low bleed characteristics which make it suitable for combined gas chromatography–mass spectrometry (GC–MS). This is particularly important when cytokinin glucosides are being examined, as high column temperatures (up to 300 °C) may have to be employed.

The major problems encountered in the GLC analysis of cytokinins relate to the preparation of the TMSi derivatives. *Bis*-trimethylsilyl-acetamide (BSA) or *bis*-trimethylsilyltrifluoroacetamide (BSTFA) in

(a)

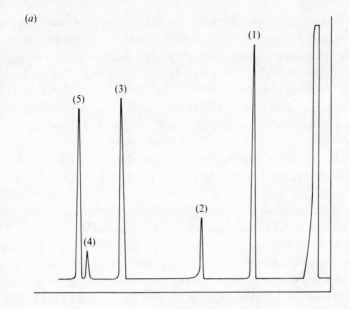

(b)

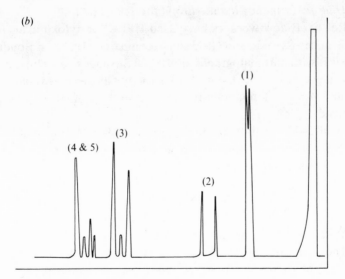

Fig. 4. GLC on 2% OV-1 columns ($1.5 \times 4 \times 10^{-3}$ m internal diameter) of TMSi derivatives (1) N^6-(Δ^2-*iso*-pentenyl)adenine; (2) zeatin; (3) N^6-(Δ^2-*iso*-pentenyl)adenosine; (4) *cis*-zeatin riboside; (5) *trans*-zeatin riboside.

N_2 flow rate = 40 cm³ min⁻¹. Temperature programme 180–300 °C at 12 °C min⁻¹.

(a) Dry sample; (b) sample containing traces of moisture.

various solvents (pyridine, dimethyl formamide, acetonitrole) have been used to prepare these derivatives. In the author's experience the problems of poor or irreproducible derivatives can be traced to impure reagents or traces of moisture in the solvents or samples. Samples must be dried rigorously, preferably at 10^{-2} mm Hg over P_2O_5 for 24 h. Reagents should be purchased in sealed ampoules from a reputable supplier (e.g. Pierce Chemical Company). The author has found that heating cytokinin samples with BSA alone at 80 °C for 1 h in a sealed tube gives optimum yields of TMSi derivatives and reproducible degrees of derivatisation. Fig. 4 shows the sorts of problems that may be encountered with samples that are not completely dry.

It should be noted at this point that certain cytokinins do not appear to be amenable to GLC analysis. The author has found that whilst TMSi zeatin-7- and 9-β-D-glucosides can be separated easily by GLC using the column previously described, TMSi zeatin- and dihydrozeatin-O-β-D-glucosides are decomposed on GLC. Workers dealing with unknown polar cytokinins should exercise care when attempting to GLC them as their TMSi derivatives.

High performance liquid chromatography (HPLC)

Relatively few workers have used HPLC for cytokinin analysis. Carnes, Brenner & Anderson (1975) have compared HPLC on Bondapak C_{18}/Porasil B with chromatography on LH-20 Sephadex as techniques for analysing the cytokinin content of tomato root pressure exudate. The compounds with cytokinin activity detected in the column eluates were not identified.

HPLC will undoubtedly become a major tool for cytokinin analysis but at the present stage it is not possible to assess its potential fully, since the purity of fractions containing cytokinin activity eluting from HPLC columns has not yet been rigorously determined.

Methods of identification of cytokinins

The small quantities of cytokinins obtainable from most plant sources limit the use of available spectroscopic techniques to ultraviolet (UV) and mass spectrometry (MS). Fortunately these two techniques are both usable with microgram quantities of material and can frequently provide enough information on an unknown cytokinin to enable a sufficiently accurate determination of its structure to be made to warrant attempts at unambiguous identification by chemical synthesis.

UV spectroscopy

When attempting to elucidate the structure of an unknown cyto-kinin the determination of its UV spectra at acid, neutral and basic pH is essential as a guide to the pattern of substitution of the purine nucleus. Details of the technique of interpreting UV spectra of cytokinins is beyond the scope of this work. However, in this connection the paper by Leonard, Carraway & Helgeson (1965) will provide detailed information.

Mass spectrometry

The mass spectra of purinyl cytokinins is usually of considerable significance and readers are referred to papers by Shannon & Letham (1966) and Hecht, Gupta & Leonard (1970) for a full treatment of this subject.

GC–MS

TMSi cytokinins are readily amenable to GC–MS analysis and will pass through most of the commercially available molecular separators. The author has found that even TMSi cytokinins which require column temperatures as high as 300 °C for satisfactory elution will pass through a silicone rubber membrane separator operating at 250 °C. Some peak broadening is observable but this is usually not serious.

Unfortunately, however, in contrast to the mass spectra of the free compounds the mass spectra of TMSi cytokinins are not very diagnostic of structure. The TMSi derivatives provide good fingerprint spectra for comparative purposes but much of the information required for the structure determination of unknown compounds is lost because of the influence of the TMSi group on the fragmentation.

When GC–MS analysis is attempted on plant extracts showing a complex of GC peaks it may be impossible to tell from the mass spectra obtained which GC peak(s) represents the active molecule(s). As a solution to this problem the author advocates the use of preparative GLC to locate the peaks in the complex which contains biological activity. GC–MS analyses may then be undertaken with some hope at least of finding out which spectra belong to the active compounds.

In the author's view the use of GC–MS can be very helpful in the elucidation of the structure of unknown cytokinins because it can provide an excellent guide to the purity of the samples to be investigated by UV spectroscopy and direct insertion MS. Before any value can be given to the interpretation of the UV spectrum or mass spectrum of an unknown compound it is essential to know that it is pure or at least to have some idea of the nature of any impurities present. For example, if preparative

GLC reveals the presence of cytokinin activity associated with a particular peak on a GLC trace, repetitive scanning GC–MS can be used to assess the purity of the peak by taking spectra at several points on the peak. If these spectra are identical, even though they may give little structural information, the compound may be isolated by preparative GLC, the TMSi group removed by mild hydrolysis, and the rest of the structure may be determined by conventional MS and UV analysis. An example of the use of this approach to determine the structure of an unknown cytokinin can be found in the paper by Horgan *et al.* (1975).

Enzymatic methods

Little work has been done on techniques of isolating and identifying cytokinin nucleotides as such. In general enzymatic degradation to the corresponding nucleoside and identification of this is the method of choice. The enzyme most frequently used for such degradations has been alkaline phosphatase obtained from a variety of sources.

Tentative identification of cytokinin glucosides has often been claimed on the basis of changing chromatographic properties of biologically active peaks on treatment with glucosidases. The author does not think it necessary to comment on the above methods except to urge other workers to use enzymes of the highest possible purity, since impure alkaline phosphatase preparations can contain glucosidase activity and vice versa.

Future developments

As has been mentioned previously, HPLC will certainly become one of the major methods of cytokinin analysis. It is hoped that workers using this technique will use discretion when claiming that a particular peak observed by a UV detector belongs to a specific cytokinin. In some ways it is unfortunate that the sensitivity of UV detectors for HPLC use is so high. The fact that they can easily detect nanogram quantities of cytokinins means that considerable scale-up will usually be necessary if peaks that are claimed to be cytokinins are to be analysed by rigorous means; e.g. UV spectra, MS and GC–MS. This, however, will have to be done particularly if the results of measurements of cytokinin levels made by HPLC are to have any meaning.

The development of multiple ion monitoring by high resolution MS will be a technique that should be of great use to workers attempting to measure the quantities of cytokinins in relatively impure extracts and may be of equal utility for assessing the purity of HPLC peaks.

The technique of field desorption MS with its ability to obtain molecular ions from very involatile compounds will probably be used more to obtain molecular weights of the presently unknown very polar cytokinins.

In conclusion, it is hoped that this brief review, incomplete as it is, will be of use to other researchers in the field in pointing out some of the problems associated with isolating and identifying cytokinins and in suggesting some methods and sources of information to help overcome these problems.

The author wishes to thank all his colleagues at Aberystwyth who have contributed to the store of knowledge that he has drawn on for this article.

Particular thanks are due to Dr J. M. Horgan, Dr J. G. Purse, Dr R. Menhenett, Mr T. L. Wang, Mr W. Davies, Miss J. E. Halfhide, Dr R. Lorenzi, Mr A. G. Thompson and Dr T. Stutchbury.

References

Armstrong, D. J., Burrows, W. J., Evans, P. K. & Skoog, F. (1969). Isolation of cytokinins from tRNA. *Biochimica et Biophysica Acta*, 37, 451.

Biddington, N. L. & Thomas, T. H. (1973). Chromatography of five cytokinins on an insoluble polyvinylpyrrolidone column. *Journal of Chromatography*, 75, 122–3.

Bieleski, R. L. (1964). The problem of halting enzyme action when extracting plant tissues. *Analytical Biochemistry*, 9, 431–42.

Burrows, W. J., Armstrong, D. J., Kammek, M., Skoog, F., Bock, R. M., Hecht, S. M., Dammann, L. G., Leonard, N. J. & Occolowitz, J. (1970). Isolation and identification of four cytokinins from wheat germ transfer RNA. *Biochemistry*, 9, 1867–72.

Carnes, M. G., Brenner, M. L. & Anderson, C. R. (1975). Comparison of reversed-phase high-pressure liquid chromatography with Sephadex LH-20 for cytokinin analysis of tomato root pressure exudate. *Journal of Chromatography*, 108, 95–106.

Dyson, W. H. & Hall, R. H. (1972). N^6-(Δ^2-isopentenyl)adenosine: its occurrence as a free nucleoside in an autonomous strain of tobacco tissue. *Plant Physiology*, 50, 616–21.

Glenn, J. L., Kuo, C. C., Durley, R. C. & Pharis, R. P. (1972). Use of insoluble polyvinylpyrrolidone for purification of plant extracts and chromatography of plant hormones. *Phytochemistry*, 11, 345–51.

Hecht, S. M., Gupta, A. S. & Leonard, N. J. (1970). Mass spectra of ribonucleoside components of tRNA. II. *Analytical Biochemistry*, 38, 230–51.

Horgan, R., Hewett, E. W., Horgan, J. M., Purse, J. & Wareing, P. F. (1975). A new cytokinin from *Populus×Robusta*. *Phytochemistry*, 14, 1005–8.

Laloue, M., Terrine, C. & Gawer, M. (1974). Cytokinins: formation of the nucleoside-5′-triphosphate in tobacco and *Acer* cells. *FEBS Letters*, 46, 45–9.

Laloue, M., Gawer, M. & Terrine, C. (1975). Modalités de l'utilisation des cytokinines exogènes par les cellules de tabac cultivées en milieu liquide agité. *Physiologie végétale*, **13**, 781–96.

Leonard, N., Carraway, K. & Helgeson, J. (1965). Characterisation of N_x,N_y-disubstituted adenines by ultraviolet absorption spectra. *Journal of Heterocyclic Chemistry*, **2**, 291–7.

Letham, D. S. (1973). Cytokinins from *Zea mays*. *Phytochemistry*, **12**, 2445.

Letham, D. S. (1974). Regulators of cell division in plant tissues. XXI. Distribution coefficients for cytokinins. *Planta, Berlin*, **118**, 361–4.

Letham, D. S., Wilson, M. M., Parker, C. W., Jenkins, I. D., Macleod, J. K. & Simmons, R. E. (1975). Regulators of cell division in plant tissues. XXIII. The identity of an unusual metabolite of 6-benzylaminopurine. *Biochimica et biophysica acta*, **399**, 61–70.

Parker, C. W. & Letham, D. S. (1973a). Regulators of cell division in plant tissues. XVI. Metabolism of zeatin by radish cotyledons and hypocotyls. *Planta, Berlin*, **114**, 199–218.

Parker, C. W. & Letham, D. S. (1973b). Regulators of cell division in plant tissues. XVIII. Metabolism of zeatin in *Zea mays* seedlings. *Planta, Berlin*, **115**, 337–44.

Playtis, A. J. & Leonard, N. J. (1971). The synthesis of ribosyl-cis-zeatin and thin layer chromatographic separation of the *cis* and *trans* isomers of ribosyl zeatin. *Biochemical and Biophysical Research Communications*, **45**, 1–5.

Purse, J. G. (1974). Studies on endogenous cytokinins in plants. Ph.D. Thesis, University of Wales.

Shannon, J. S. & Letham, D. S. (1966). Regulators of cell division in plant tissues. IV. The mass spectra of cytokinins and other 6-aminopurines. *New Zealand Journal of Science*, **9**, 833–42.

Vreman, H. J. & Corse, J. (1975). Recovery of cytokinins from cation exchange resins. *Physiologia plantarum*, **35**, 333–6.

Wang, T. L., Thompson, A. G. & Horgan, R. (1977). A cytokinin glucoside from the leaves of *Phaseolus vulgaris* L. *Planta, Berlin*, **132**, 285–8.

P. F. SAUNDERS

The identification and quantitative analysis of abscisic acid in plant extracts

Introduction

The term 'inhibitor β', first used by Bennet-Clarke & Kefford (1953), does not refer to a specific compound but rather to a zone of growth inhibitory activity found at a particular R_f when partially fractionated plant extracts are chromatographed on paper. The zone is almost invariably found when an acidic, ether-soluble fraction is prepared from almost any higher plant tissue and occurs at about R_f 0.6–0.7 on chromatograms developed in propan-2-ol:ammonia:water (PAW) (10:1:1::v:v:v). Abscisic acid (ABA), which was originally known as 'abscisin II' or 'dormin', was first identified in extracts of sycamore leaves (Cornforth, Milborrow, Ryback & Wareing, 1965) and of cotton fruits (Ohkuma, Addicott, Smith & Thiessen, 1965). This compound has strong inhibitory activity in the bioassays used to detect inhibitor β and also has an R_f of 0.6–0.7 when chromatographed on paper in PAW. Moreover, ABA is found in the acidic, ether-soluble fraction of extracts from a wide range of plant tissues: in my own laboratory, for example, we have succeeded in identifying it in almost every tissue we have examined, ranging from the arils of *Taxus baccata* to the roots of *Salix viminalis* and the leaves of *Raphanus sativus*. On these grounds, there can be little doubt that the ABA content of plant extracts contributes, in large measure, to the biological activity of the inhibitor β zone. With several tissue extracts, Milborrow (1968) showed by direct measurement that the ABA content was sufficient to account entirely for the observed inhibitor β activity. It cannot be too strongly emphasised, however, that inhibitor β and ABA are not synonymous.

All too frequently, especially in literature reviews, reference is made to observations on the ABA content of tissues, when what was actually measured was inhibitor β activity. Indeed, it sometimes happens that the observations referred to were made long before ABA was discovered. The danger of such unwarranted extrapolations is illustrated by published data of Alvim, Hewett & Saunders (1976) which show the inhibitor β activity of fractions derived from willow xylem sap collected at various times of year. Comparison of the data for 30 April and 30 June, for example, indicate an increase in inhibitor β activity. Direct measurement of ABA content, however, showed that the concentration in the sap declined from 20 μg dm^{-3} on 30 April to 15 μg dm^{-3} on 30 June. Conversely, we have sometimes observed quite large differences in the ABA content of extracts which were not paralleled by significant differences in inhibitor β activity. Aside from the inherent variability of bioassay methods, several possible reasons can be advanced for discrepancies between measurements of inhibitor β activity and of ABA content. First, unless care is taken in selecting appropriate concentrations of test solutions, bioassays can be extremely insensitive. Bioassays based on the inhibition of extension growth in wheat coleoptile sections, for example, show a more or less logarithmic dose–response curve. Accordingly, such bioassays are effectively saturated and incapable of revealing differences in inhibitory activity at concentrations of ABA which do not completely inhibit extension growth. This particular shortcoming of bioassays is widely recognised, although often ignored, and with some extra effort can be circumvented by assaying serial dilutions of the samples to be compared.

A much more troublesome problem concerns the possibility of interference in the bioassay by other components of a plant extract which may not be separated from ABA before inhibitor β activity is determined. Such interference could conceivably derive from the presence of other inhibitory compounds, or possibly from the presence of compounds which are not inhibitory *per se* but which have a synergistic effect on the response of the bioassay to ABA. Alternatively, the inhibitor β zone may contain growth promoting compounds or compounds which can overcome the inhibitory effect of ABA. Some indication of the possibilities is given by Table 1 which shows the R_f of a number of compounds in PAW. Amongst compounds which could conceivably contribute to the bioassay activity of the inhibitor β zone along with ABA (R_f 0.50–0.72) are salicylic acid (R_f 0.53–0.62), indole acetamide (R_f 0.68–0.80) and several gibberellins. Although salicylic acid is a much less potent growth inhibitor than ABA, it is likely to occur in some plant extracts (notably of *Salix*) in far greater quantities. Indole acetamide, which could obscure the effect of

Table 1. *The R_f values of various compounds on paper developed in propan-2-ol: 0.880 ammonia: water (10: 1: 1:: v: v: v) (Lenton, unpublished)*

Salicylic acid	0.53–0.63	Gibberellin A$_7$	0.56–0.70
Cinnamic acid	0.48–0.65	Gibberellin A$_9$	0.60–0.70
Indole acetamide	0.68–0.80	Abscisic acid	0.56–0.72
Gibberellin A$_4$	0·56–0·70	2,*trans*-abscisic acid	0.56–0.72
Gibberellin A$_5$	0·50–0·60		

ABA by virtue of its auxin activity, is known to be formed from naturally occurring sugar esters of IAA during chromatography in ammoniacal solvents (Klämbt, 1961). Gibberellins do not normally give a response in the wheat coleoptile bioassay but some can partially counteract inhibition by ABA.

To some extent, the likelihood of interference by other compounds can be lessened by using several different bioassays. The bioassay devised by Ogunkanmi, Tucker & Mansfield (1973), for example, which is based on the effect of ABA on stomatal closure appears to show a high degree of specificity. Alternatively, steps can be taken to separate ABA from any identifiable interfering substances which are known to accompany ABA in the inhibitor β zone. Generally speaking, however, ABA represents a very small proportion by weight of the inhibitor β fraction and, since the identity of possible interfering substances is to a large extent unknown, any quantitative results based on bioassays must be subject to some degree of uncertainty. Despite the need for caution, if two plant extracts show a difference in inhibitor β activity, it is reasonable to suggest that the ABA content is different: to this extent bioassays provide a useful first approach when studying the effect of various treatments on ABA levels. Clearly, however, any conclusions drawn must remain tentative pending direct measurement of ABA levels by methods showing greater specificity. Since its discovery in 1965 several methods for the identification and quantitative analysis of ABA have been developed based on the physical and chemical properties of the molecule rather than on its biological effects. Some methods involve considerable investment in capital equipment, some are more reliable than others, none is foolproof. The following account constitutes a review and personal assessment of some of the methods available, with considerable emphasis on the methods which appear to afford the best compromise between cost, convenience and reliability.

Identification of ABA on the basis of various physical and chemical properties

Optical activity

Naturally occurring ABA is the R(+) enantiomer. The compound absorbs strongly in the UV ($E_{260} = 22\,000$) and because of the asymmetric molecular environment of the chromophores, the optical rotatory dispersion (ORD) curve shows a pronounced anomaly in the region of the UV absorption maximum consisting of two overlapping Cotton effects. Cornforth, Milborrow & Ryback (1966) have published the following characteristics for the ORD in acidified methanol of R(+)-ABA purified from sycamore leaves:

nm	$[\alpha]$
589	+430
289	+24000
269	0
246	−69000
225	0

The shape of the ORD curve for ABA in acidified methanol is quite characteristic and affords a useful means of confirming the presence of the compound in a purified extract. Because of the extremely high specific rotation at the extrema of the Cotton anomaly, ORD can be determined in quite dilute samples: a solution containing 1 μg cm^{-3} in a 1 cm cuvette will produce an optical rotation of about +2.5 millidegrees at 289 nm – well within the range of modern spectropolarimeters. Measurement is possible, however, only in the absence of appreciable light absorption by other compounds and samples must have an OD of no more than 1.0 over the wavelength range of interest. Furthermore, distortion of the ORD curve by optically active contaminants is sometimes apparent with samples purified from plant extracts. For these reasons, quite extensive purification is frequently required before satisfactory curves can be obtained. The curve shown in Fig. 1 was obtained with a sample purified from 120 cm^3 of birch spring sap by treatment with polyvinylpyrrolidone (PVP) followed by thin-layer chromatography (TLC) of an acidic, ether-soluble fraction. The features of this curve correspond closely with those of pure R(+)-ABA. Many plant extracts will, however, require considerably more purification.

Spectropolarimetry can be used to determine the concentration of ABA in purified fractions, provided that the sample is free from significant

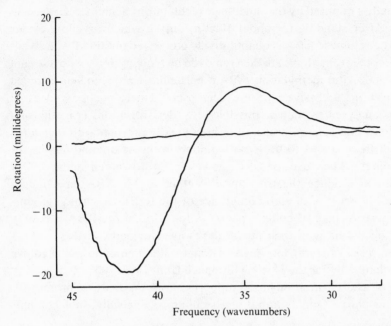

Fig. 1. Optical rotatory dispersion curve of a sample of ABA purified from 120 cm³ birch xylem sap. Sap was treated with PVP at pH 7.0, extracted with ether at pH 3.5 and the ether-soluble fraction subjected to TLC on silica gel in toluene:ethyl acetate:acetic acid (40:5:2::v:v:v). ORD was determined in acidified ethanol using 1 cm cells in a Bellingham and Stanley Polarmatic 62 spectropolarimeter (Harrison & Saunders, unpublished).

contamination by other optically active compounds whose presence will usually be apparent from their effect on the precise shape of the curve. Unfortunately it is difficult to obtain an exact figure for specific rotation. My own measurements on an extensively purified sample from lemon juice in which the ABA concentration was determined by gas–liquid chromatography (GLC) yielded values of $+15\,500$ for $[\alpha]_{289}$ and of $-55\,500$ for $[\alpha]_{245}$. In practice, of course, relative concentrations are of more interest than absolute amounts and slight discrepancies in published values for specific rotation are usually of little consequence.

Gas chromatographic properties

Free ABA is not sufficiently volatile to be analysed by GLC but is readily converted to its methyl ester (Me-ABA) which can readily be chromatographed using a wide range of stationary phases. Derivatisation is best accomplished by dissolving the sample in a mixture of acetone and methanol (9:1::v:v) and bubbling through it a stream of diazomethane generated as required by the small scale technique of Schlenk & Gellerman (1960). After standing for about 10 min, during which a yellow colour should be retained, ABA is quantitatively converted to Me-ABA. Solvent and excess diazomethane are then removed by blowing down with a stream of nitrogen. After methylation, ABA is particularly prone to light-induced isomerisation. Excessive exposure to light should therefore be avoided and samples analysed as soon as possible after derivatisation. Trimethylsilyl derivatives can also be prepared in high yield with standard procedures although there seems to be some variation in the nature and number of the products. Their use for GLC, however, affords no advantages. Our attempts to produce the trimethylsilyl ether of Me-ABA have failed, presumably because of steric hindrance of the tertiary hydroxyl group.

The performance of column packings for GLC of Me-ABA is readily judged on the basis of their ability to resolve the methyl esters of ABA and its 2,*trans* isomer (*trans*-ABA). Of many stationary phases tested, we have obtained by far the best performance from Epon 1001 used as a 2% loading on a 80/100 mesh silanised support such as AW-DMCS Chromosorb W or Gas-chrom Q in 3 mm i.d. glass columns. Typically, such columns

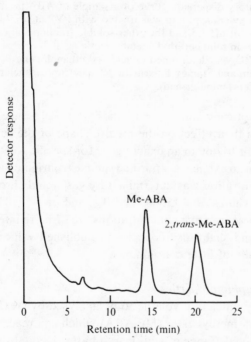

Fig. 2. Separation of the methyl esters of ABA and its 2,*trans* geometrical isomer by GLC on 2% Epon 1001 using FID (Lenton & Saunders, unpublished).

will attain efficiencies of up to 20 theoretical plates cm⁻¹ for Me-ABA and provide excellent resolution of Me-ABA and *trans*-Me-ABA with little or no tailing (Fig. 2). For routine analysis we have used 150 cm columns at 210 °C with a carrier gas flow rate of 40 cm³ min⁻¹; under these conditions Me-ABA has a retention time of about 15 min. Better resolution and efficiency can be achieved at lower carrier flow rates but at the expense of unnecessarily long analysis times. The silicone grease stationary phases OV-1 and OV-17 are also suitable for GLC of Me-ABA and, because of the exceptionally low column bleed, are especially useful in applications involving combined gas chromatography–mass spectrometry (GC–MS). A 2% loading of OV-1 on 100–120 mesh Gaschrom Q in 150 cm columns operated at 200 °C with a carrier flow rate of 40 cm³ min⁻¹ separates Me-ABA, with a retention time of about 4 min.

Although Me-ABA will produce a symmetrical peak with a characteristic retention time using stationary phases such as Epon 1001, it is not safe to assume that a corresponding peak in the gas chromatogram of a partially purified and methylated plant extract is also caused by Me-ABA. Even co-chromatography with an authentic sample of Me-ABA is insufficient to confirm the identity of such a peak. Fortunately, a very simple test can be used to confirm peak identity. Abscisic acid, especially when esterified, is readily converted to its 2,*trans* geometrical isomer which does not appear to occur naturally unless in very small amounts. Isomerisation

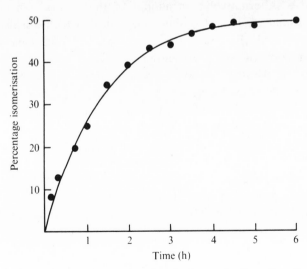

Fig. 3. Time-course for the isomerisation of methyl abscisate during exposure of a solution containing 100 mg dm⁻³ to light from a Hanovia Model 16 UV lamp. (Redrawn, with permission, from Lenton, Perry & Saunders, 1971.)

is induced by UV light, irradiation leading to the establishment of an equilibrium mixture containing approximately equal amounts of the two isomers. Fig. 3 shows the time-course for the isomerisation of an authentic sample of Me-ABA (100 mg dm^{-3} in methanol) placed in a silica cuvette in front of a Hanovia Model 16 far-UV light source. More rapid isomerisation is produced by irradiation with light having a greater intensity at the UV absorption maximum of ABA. The 254 nm source in a DeSaga Miniuvis lamp, for example, can cause 20% isomerisation in about 5 min. Depending on the light source a certain amount of irreversible degradation to other products also occurs. After 3 h in front of the DeSaga lamp, for example, all traces of Me-ABA and *trans*-Me-ABA had disappeared.

To confirm the identity of a presumptive Me-ABA peak on a gas chromatogram any remaining sample is irradiated alongside an authentic sample of Me-ABA and an aliquot is reinjected. Fig. 4 shows chromatograms of a partially purified sample from birch buds before and after UV-induced isomerisation. Irradiation has led to a reduction in the size of the presumptive Me-ABA peak and in the simultaneous appearance of a peak corresponding in retention time with *trans*-Me-ABA. An authentic sample of Me-ABA behaved in an identical fashion. Confirmation that the Me-ABA peak does not include a contribution from other components can be achieved by further irradiation or by the use of a more intense UV source which will lead to complete disappearance of a peak which is caused entirely by Me-ABA. Alternatively, an aliquot of the sample can be reduced by adding a small amount of sodium borohydride which will also eliminate an Me-ABA peak. Final confirmation of peak identity can, if necessary, be obtained by GC–MS (see below).

Although the UV-isomerisation and sodium borohydride reduction tests are extremely useful, they must be used with care. Unless an absolutely reproducible source of UV radiation is available, an authentic Me-ABA sample should be tested at the same time. Even then, identical behaviour of Me-ABA in a partially purified plant extract cannot be assumed if the sample is not sufficiently free from other UV-absorbing materials which could conceivably either reduce or enhance the rate of isomerisation. Despite the need for caution, however, the technique will readily distinguish between ABA and various fatty acids, plasticisers and paraffins which can easily be mistaken for ABA if retention time is used as the sole criterion of identity.

Having obtained satisfactory evidence that a peak is caused entirely by Me-ABA, gas chromatography can readily be used for quantitative analysis of ABA in purified extracts. A flame-ionisation detector (FID) will produce a linear response over a very wide range, ultimate sensitivity being

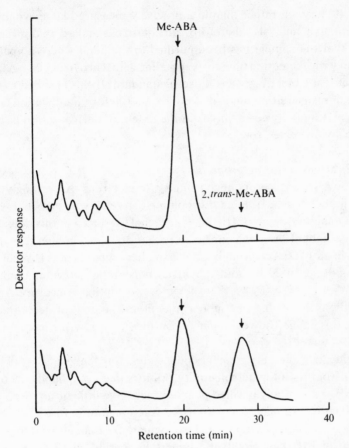

Fig. 4. Gas chromatograms of a purified, methylated extract from birch buds before and after isomerisation by UV light. Methanolic extracts were purified by isolating an ether-soluble, acidic fraction which was then subjected to column chromatography on PVP followed by TLC. Samples equivalent to 0.025 g fresh weight of tissue were chromatographed on 2% Epon 1001 using ECD. (Redrawn, with permission, from Harrison & Saunders, 1975.)

restricted primarily by the signal: noise ratio of the associated electronics. Provided that the chromatogram has a satisfactory base line, peak area measurement affords the best means of quantification and for most purposes can be achieved with sufficient accuracy by tracing and weighing. With an F & M 402 gas chromatograph we have used an FID to measure ABA levels in injected samples containing down to 0.01 μg in 5 mm^3. Best results are achieved, however, if 1 mm^3 samples containing 0.2–1 μg mm^{-3} can be injected. With samples purified from plant extracts it is often convenient to dissolve in a minimal arbitrary volume of acetone and methanol (8:1::v:v), rather than in a precisely measured larger volume. For quantitative analysis, therefore, an internal standard is desirable. Provided that the sample has been purified to the point where no peaks are present with a retention time corresponding with that of *trans*-Me-ABA, the geometrical isomer provides a useful standard. If this is added to the sample as a measured amount of the free acid before methylation, any possible variation between samples in the extent of derivatisation is also automatically corrected for.

Electron capturing properties

Me-ABA has a pronounced affinity for electrons. Accordingly, its presence in the effluent from a GLC column can readily be detected with an electron-capture detector (ECD). The signal produced in this detector by a given weight of Me-ABA can be more than 100 times greater than that produced in an FID. Accordingly, much smaller amounts of ABA can be detected and measured without the signal:noise ratio becoming unmanageable. With the F & M 402 ^{63}Ni detector we routinely measure ABA levels of the order of 0.1 to 5 ng per injection. The limit of detection is far below this range, however, and quantities of a few picograms can readily be detected.

In practice, considerably greater benefit stems from the selectivity of the ECD than from its ultimate sensitivity. Many other compounds accompanying ABA in partially purified plant extracts give little or no response in the ECD. Accordingly, less extensive fractionation before GLC can frequently be tolerated. Often, analyses which are difficult or impracticable with an FID can readily be undertaken using an ECD. Results obtained with a particularly 'dirty' preparation from birch buds are shown in Fig. 5. With the FID, Me-ABA appears as a barely visible shoulder on a long pyrolysis tail. By contrast, the ECD produced a large symmetrical peak on a flat base line from a much smaller sample. When the selectivity of the ECD is relied upon to permit detection of Me-ABA in the presence of other material having the same retention time, it is obviously impossible to obtain absolute confirmation of peak identity by GC–MS. In these circumstances, the UV-isomerisation test described above becomes indispensable.

The greater sensitivity of the ECD permits quantitative analyses of ABA in much smaller samples of plant material than is possible with an FID. Under the particular operating conditions used routinely in my laboratory, a linear response in terms of peak area is obtained with injections containing up to 20 ng ABA. It should be remembered, however, that the ECD is saturable and that response will begin to depart from linearity when an

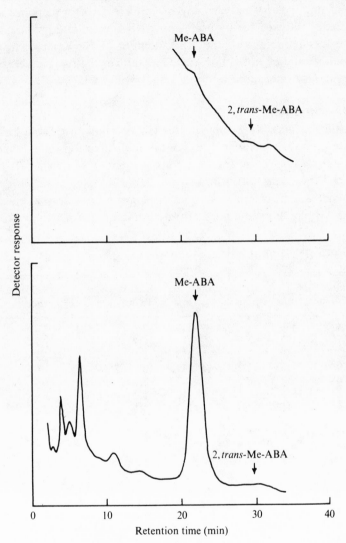

Fig. 5. Gas chromatograms of the partially purified, methylated extract of birch buds obtained using (a) flame-ionisation detector (b) electron capture detector. Samples injected were equivalent to (a) 0.25 g and (b) 0.025 g fresh weight of tissue (Harrison & Saunders, unpublished).

appreciable proportion of the detector standing current has been abolished by electron capture. Under our usual operating conditions, the peak height produced from a 20 ng injection represents an approximately 25% reduction in standing current. Samples containing sufficient ABA to cause a greater reduction in standing current are always diluted and re-run. Minor departures from linearity and also any variation in detector response with time are readily allowed for by using *trans*-ABA as an internal standard in amounts comparable with the ABA content of the sample. In this connection it should be noted that equal amounts of the two isomers give an equal response, measured by peak area, in the ECD.

Mass spectrum

A typical mass spectrum of Me-ABA is shown in Table 2. The fragmentation pattern includes a very weak molecular ion (m/e 278) but abundant and characteristic fragments are present at m/e 190, 162, 134, 125, and 91. The mass spectrum of *trans*-Me-ABA is similar, differing chiefly in that the fragment at m/e 125 is considerably less abundant. Samples of ABA of sufficient purity for determination of mass spectra by

Table 2. *Mass spectrum of methyl abscisate*

m/e	Relative abundance	m/e	Relative abundance	m/e	Relative abundance	m/e	Relative abundance
260	2.1	159	1.6	117	2.2	79	3.4
246	2.7	152	1.9	115	1.2	78	1.2
245	1.2	149	2.3	113	1.5	77	1.7
231	1.0	148	2.2	112	10.7	73	1.0
222	1.4	147	6.7	111	2.7	69	5.5
205	2.8	146	3.2	110	1.2	68	4.9
204	1.1	145	3.4	109	2.5	67	8.4
203	1.4	141	1.7	108	2.3	66	2.1
201	1.0	139	1.8	107	5.9	65	1.9
192	1.6	138	1.2	106	7.0	59	2.7
191	14.1	137	1.1	105	4.5	56	3.2
190	100.0	136	3.1	98	1.0	55	7.6
189	2.2	135	11.0	97	3.4	54	1.4
187	2.1	134	21.3	96	9.2	53	1.6
176	1.2	133	5.1	95	5.2	44	9.8
175	2.5	131	1.3	94	3.4	43	6.5
172	1.8	126	2.5	93	3.8	42	1.8
165	3.4	125	35.4	92	1.1	41	7.6
164	1.2	123	2.4	91	10.4	39	3.5
163	4.5	122	3.1	85	1.3	31	1.7
162	22.1	121	3.8	83	7.9	29	7.8
161	8.8	120	1.8	82	2.1	28	1.0
160	2.0	119	5.5				

These data were obtained by GC–MS using an AEI MS30 mass spectrometer fitted with a membrane separator, operating at 24 eV with a scanning rate of 3 s decade^{-1}. Fragments with a relative abundance less than 1% have been omitted. (By courtesy of Mr J. K. Heald.)

direct probe analysis can be obtained from plant material only with difficulty. The use of GC–MS, however, readily permits the identification of Me-ABA in methylated fractions from plant extracts which have been purified to the extent normally undertaken with a view to analysis by GLC using an FID. Acceptable mass spectra can usually be obtained from injected samples containing 0.1–1 μg ABA (e.g. Most, Gaskin & Mac-Millan, 1970).

In Aberystwyth we have used GC–MS to obtain mass spectra from samples from a wide range of plant materials in which gas chromatogram peaks had been identified as Me-ABA by application of the UV-isomerisation test. In all cases, mass spectra have confirmed that the peaks were indeed caused almost entirely by Me-ABA. Samples prepared for analysis using the ECD are not necessarily sufficiently pure nor sufficiently rich in ABA for GC–MS. By extrapolation, however, we are confident that peaks identified by UV-isomerisation contain no other components which contribute significantly to the ECD signal.

By focusing at a particular m/e value it is possible to use a mass spectrometer as a highly specific GC detector (e.g. Railton, Reid, Gaskin & MacMillan, 1974). For analysis of samples containing Me-ABA it is convenient to monitor the ion current at m/e 190. The appearance of a peak at this m/e value which corresponds in retention time with the authentic compound, can leave little doubt that the peak is caused entirely by the presence of Me-ABA.

When used for single ion current monitoring a mass spectrometer can be operated satisfactorily at much greater sensitivities than are possible in a scanning mode. Accordingly, although microgram quantities are required for GC–MS, samples containing 1 ng or less can easily be detected by single ion current monitoring.

Purification of ABA from plant extracts

The extent to which a plant extract must be purified prior to analysis will depend on the interaction of a number of factors including the analytical procedure to be used, the amount of ABA present and the nature and amount of other compounds in the extract. It is clearly impossible, therefore, to give precise instructions which will be generally applicable. Outlines of the purification procedures used to obtain some of the data presented in this paper are given in the legends to the figures. It must be emphasised, however, that these procedures could well be inadequate for some extracts and perhaps unnecessarily extensive for others. Table 3 shows a generalised procedure which we would take as

Table 3. *Outline of generalised procedure for purification of ABA from methanolic extracts*

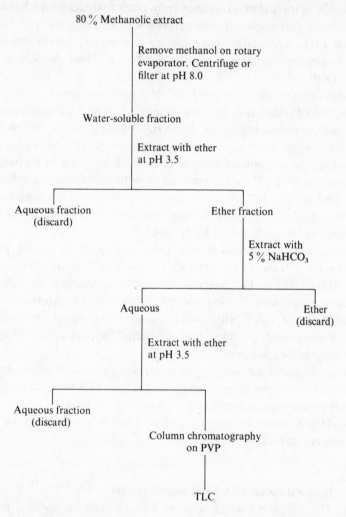

a starting point in developing a technique for the pre-purification of ABA from an 80% methanol extract of an unfamiliar tissue regardless of which analytical procedure we ultimately decided to adopt. In the light of experience, this procedure would almost certainly be either condensed by omitting or simplifying certain steps or extended by, for example, incorporating further TLC steps.

Preparation of water-soluble fraction

Removal of methanol is best accomplished in a rotary evaporator at about 35 °C to leave an aqueous residue. With most extracts it is desirable to add sufficient alkali to the crude extract to maintain an approximately neutral pH in the final aqueous solution. After evaporation, the pH should be adjusted to about 8.0. These pH adjustments serve both to minimise isomerisation of ABA to *trans*-ABA, which occurs most readily in acidic solutions, and more importantly to ensure that ABA is ionised in which state it is less likely to be lost by co-precipitation with water-insoluble materials. Omission of this precaution often leads to very much reduced yields of ABA. Insoluble material in the aqueous residue can be removed by filtration or centrifugation. At this stage, clarification is often greatly assisted by freezing and thawing the extract to coagulate emulsified lipid.

Preparation of ether-soluble acidic fraction

At pH 3.0–3.5, ABA is undissociated and the partition coefficient between water and diethyl ether strongly favours the organic phase. After pH adjustment, three extractions with equal volumes of ether are sufficient to give almost complete recovery of ABA. The resultant ether solution also contains non-acidic substances and ABA can to a large extent be separated from these by partitioning with 5 % sodium bicarbonate solution. Acidic substances, including ABA, can then be recovered by adjusting the bicarbonate extract to pH 3.5 and extracting with three equal volumes of ether. The bulked ether extracts are then dried with anhydrous sodium sulphate and evaporated.

Treatment with polyvinylpyrrolidone (PVP)

The ether-soluble acidic fraction from most plant extracts will require further purification before TLC unless very small samples are sufficient for the intended analysis. A very considerable reduction in dry weight can be achieved at this stage, with very little loss of ABA, by chromatography on a column of insoluble PVP. The commercial product Polyclar AT is suitable but should be washed and fines should be removed by several cycles of suspension in distilled water. Columns are packed and run under gravity in 0.1 mol dm^{-3} phosphate buffer at pH 6.0. Samples are loaded as an aqueous solution which can usually be readily prepared by dissolving in a little 0.1 mol dm^{-3} K_2HPO_4 solution followed by pH adjustment to 6.0. With some samples it is easier to prepare the ammonium salt of the ether-soluble acidic fraction by adding a little 0.880 ammonia solution, removing excess ammonia in a stream of nitrogen, and dissolving the residue in water. With a 2×23 cm column of PVP, ABA is typically eluted at pH 6.0 in a symmetrical peak between 50 and 70 cm^3. Recoveries approach 100% and, once the elution volume has been determined with authentic ABA for a given batch of PVP, identical columns are found to behave quite reproducibly. Very much smaller columns are adequate for many purposes but, since extracts from different tissues vary widely, the optimum size for a particular application is best determined by trial and error.

Often, a sufficient reduction in dry weight can be achieved by slurrying the extract with PVP and dispensing with column chromatography. Slurrying is normally carried out by adding dry, washed PVP to an aqueous solution of the extract at pH 6.0, stirring for an hour or so and filtering. On various occasions we have applied this technique to the clarified aqueous residue remaining after removal of methanol, to the bicarbonate extract, and to an aqueous solution of the ether-soluble acidic fraction.

After PVP treatment, extracts are normally adjusted to pH 3.5, extracted with ether and the ether extract dried and evaporated.

Thin-layer chromatography

Extracts purified through the stages outlined so far will usually contain ABA at up to 90% recovery. ABA will not, however, have been separated from *trans*-ABA nor from a vast range of other moderately polar organic acids normally encountered in plant extracts. For the various analytical procedures based on GLC a final step based on silica gel TLC will normally be essential for any plant extract. Except with very dirty extracts, however, or with extracts containing very low levels of ABA, no further purification beyond the TLC stage will be needed before methylation and GLC.

Because TLC will be the final step in most purifications it is essential at this stage to avoid contamination with extraneous material. Thus, all solvents should be carefully redistilled and thin layer plates should be pre-run in ethanol:acetic acid (98:2), dried and reactivated as necessary before loading the samples. Glassware should be chemically clean and all contact with plastics, Parafilm etc. should be avoided. For most purposes we have used 500 μm layers of Merck silica gel GF254 activated at 100 °C for 30 min. Loading of samples is greatly facilitated if the plates are hardened by slurrying the silica gel in water containing 0.3% carboxymethyl cellulose.

Amongst various solvent systems used we have found toluene:ethyl acetate:acetic acid (40:5:2::v:v:v) to be particularly useful. With this system, plates are usually developed to 15 cm in three successive runs.

Samples are usually line-loaded in 15 cm bands and loadings of up to 5 mg dry weight are possible. Marker spots of authentic ABA are loaded in two outer 'lanes' and, on the fluorescent plates used, are visible as dark spots under a UV lamp. During visualisation of the markers, the area carrying the extract must be carefully shielded with aluminium foil to prevent UV-induced isomerisation of ABA.

After TLC, a band corresponding with the ABA marker spots should be scraped off and eluted. For elution, the most effective solvent we have tried is water-saturated chloroform although we have more frequently used a 1:1 (v:v) mixture of acetone and methanol. Elution can be conveniently carried out through small cellulose thimbles. Whichever solvent is used for elution it is essential to elute the TLC plates when they are barely dry. Allowing them to dry in air for just a few hours can reduce recoveries from about 90% to 10% or less.

After TLC, few extracts will not be amenable to GLC. If a further thin-layer stage seems necessary, this is best carried out after methylation using ethyl acetate:hexane (1:1::v:v) as the running solvent.

Quantitative analysis

As already described, the quantitative analysis of ABA in purified extracts presents little difficulty. Depending on the amounts present and on the state of purity, quantification can be achieved by GLC using FID, ECD or single ion current monitoring or by spectropolarimetry. Determination of the amount originally present in the crude extract, however, can be reliably accomplished only if some means is available for correcting for losses during purification. Typical overall recoveries using the entire procedure outlined in Table 3 can vary from about 20% to 70%. Correction for variable recovery can be made only if an internal standard, known to be recovered in the same proportion as endogenous ABA, is added to the unfractionated extract in known amounts. The internal standard used must, of course, be capable of quantitative analysis independently of endogenous ABA after purification.

The earliest method to use an internal standard was the 'racemate dilution' method as described by Milborrow (1968). In this technique the internal standard is synthetic (RS)-ABA which is added to the unfractionated extract in a known amount. After sample purification, the amount of natural, optically active (R)(+)-ABA in the purified sample is determined spectropolarimetrically. In the method as originally described the total amount of ABA recovered was determined by measuring UV absorption at 260 nm. After subtracting the amount of optically active

ABA present, this figure was used to obtain a value for percentage recovery of the racemic ABA added to the crude extract. Since none of the steps used for purification involved optically active reagents, it could be assumed that the natural ABA present in the extract had been recovered in the same proportion and an appropriate correction could thus be made for losses during purification. When UV absorption is used to measure total ABA levels, the method involves very extensive pre-purification to ensure the complete absence of other UV-absorbing substances. The method can sometimes be applied to rather less highly purified fractions, however, if GLC is used to measure total ABA.

The geometrical isomer, 2,*trans*-ABA, was used as an internal standard to correct for losses during purification in the quantitative procedure based on GLC originally described by Lenton, Perry & Saunders (1971). Here, GLC is used to measure both the amount of endogenous ABA and the amount of added *trans*-ABA recovered in purified samples. For reliable results it is, of course, essential to check first that samples purified without added internal standard contain negligible amounts of *trans*-ABA and of other compounds with the same GLC retention time. It is also essential to check that the purification procedures used lead to equal recoveries of the two compounds. None of the purification steps described above differentiate between the geometrical isomers except for the final TLC stage which causes separation into two adjacent bands. Provided that marker spots of each isomer are run simultaneously with the extract and that a band sufficiently broad to include both compounds is eluted, equal recoveries of the two isomers are always obtained. GLC itself, of course, need not be carried out quantitatively, since the concentration of ABA in the crude extract is given, regardless of sample size, by the expression:

$$[ABA] = [trans\text{-}ABA] \times \frac{\text{Me-ABA peak area}}{trans\text{-Me-ABA peak area}}.$$

A major disadvantage of using *trans*-ABA as an internal standard is that the UV-isomerisation test can be used to confirm peak identity only on a separately purified sample containing no internal standard. This disadvantage is not present in the method which we now use routinely and in which we add ^{14}C-labelled ABA to the crude extract in order to determine recoveries. This method is basically an isotope dilution technique in which GLC, together with scintillation counting, is used to determine the specific activity of ABA in a purified fraction. Specific activity of the [^{14}C]ABA added to the crude extract is determined in the same way and the amount of endogenous ABA in the crude extract can readily be calculated from the two values. In this method, GLC must be carried out quantitatively and we usually achieve this by adding a known amount of *trans*-ABA to the purified extract as an internal standard immediately before methylation. One of the advantages of the isotope dilution technique is that it frequently makes it possible to simplify or omit some of the earlier purification steps. This is because a greater degree of purification is attained in the final TLC step by virtue of the fact that only a very narrow band need be eluted: the *trans*-ABA zone is not included and the ABA band itself need not be eluted in its entirety. Some degree of isomerisation during purification can, of course, be tolerated since this will affect the endogenous ABA and the added labelled compound equally.

Very recently, Rivier, Milon & Pilet (1977) have introduced the use of hexadeuterated ABA as an internal standard to correct for purification losses in a quantitative procedure based on single ion current monitoring. In their technique, a known amount of hexadeuterated ABA is added to the crude extract. After purification, the relative amounts of unlabelled, endogenous ABA and added, deuterated ABA are determined by gas chromatography using a mass spectrometer to monitor similtaneously the ion current at m/e 190, originating from endogenous ABA, and at m/e 194, originating from the corresponding deuterated fragment.

Conclusions

Of the methods available for quantitative analysis of ABA in plant extracts, the most generally useful techniques are those based on gas chromatography using an electron capture detector with isotope dilution to correct for recovery. The equipment required is simple to operate and relatively inexpensive. The extent of purification required is minimal and the ultimate sensitivity is at least as good as that of any other method, including bioassay, In common with all other methods, however, the technique can do no more than measure the amount of ABA in a plant extract. Like everyone else we have generally used 80% methanol to extract ABA from plant tissues. We have no means of knowing whether this solvent extracts all of the ABA: we know only that it seems to be the best extractant available. Some workers have attempted to allow for the possible inefficiency of methanol extraction by adding an internal standard during homogenisation. One then is left with the unanswerable question of whether and to what extent the internal standard equilibrates with non-extracted endogenous ABA. Even if 80% methanol does give complete extraction, our analyses give us no more than a value for the mean internal concentration of ABA in the tissue extracted. Whether such information is meaningful is, perhaps, debatable.

134   P. F. SAUNDERS

The methods discussed in this chapter have been developed over several years with the assistance of many associates. I am particularly grateful for the contributions made by Dr R. Alvim, Dr M. A. Harrison, Mr J. K. Heald, Dr R. Horgan, Dr J. R. Lenton, Mrs V. M. Perry and Dr S. Thomas. Financial assistance from the Science Research Council is gratefully acknowledged.

References

Alvim, R., Hewett, E. W. & Saunders, P. F. (1976). Seasonal variation in the hormone content of willow. I. Changes in abscisic acid content and cytokinin activity in the xylem sap. *Plant Physiology*, **57**, 474–6.

Bennet-Clarke, T. S. & Kefford, N. P. (1953). Chromatography of the growth substances in plant extracts. *Nature, London*, **171**, 645–7.

Cornforth, J. W., Milborrow, B. V. & Ryback, G. (1966). Identification and estimation of (+)-abscisin II ('dormin') in plant extracts by spectropolarimetry. *Nature, London*, **210**, 627–8.

Cornforth, J. W., Milborrow, B. V., Ryback, G. & Wareing, P. F. (1965). Identity of sycamore 'dormin' with abscisin II. *Nature, London*, **205**, 1269.

Harrison, M. A. & Saunders, P. F. (1975). The abscisic acid content of dormant birch buds. *Planta, Berlin*, **123**, 291–8.

Klämbt, H.-D. (1961). Wachstumsinduktion und Wuchsstoffmetabolismus im Weizenkoleoptilzylinder. II. Stoffwechselprodukte der Indol-3-essigsäure und der Benzoesäure. *Planta, Berlin*, **56**, 618–31.

Lenton, J. R., Perry, V. M. & Saunders, P. F. (1971). The identification and quantitative analysis of abscisic acid in plant extracts by gas liquid chromatography. *Planta, Berlin*, **96**, 271–80.

Milborrow, B. V. (1968). The identification of (+)abscisin II in plants and measurements of its concentrations. *Planta, Berlin*, **76**, 93–113.

Most, B. H., Gaskin, P. & MacMillan, J. (1970). The occurrence of abscisic acid in inhibitors B₁ and C from immature fruit of *Ceratonia siliqua*, L. (Carob) and in commercial carob syrup. *Planta, Berlin*, **92**, 41–9.

Ogunkanmi, A. B., Tucker, D. J. & Mansfield, T. A. (1973). An improved bioassay for abscisic acid and other antitranspirants. *New Phytologist*, **72**, 277–82.

Ohkuma, K., Addicott, F. T., Smith, O. E. & Thiessen, W. E. (1965). The structure of abscisin II. *Tetrahedron Letters*, **29**, 2529–35.

Railton, I. D., Reid, D. M., Gaskin, P. & MacMillan, J. (1974). Characterisation of abscisic acid in chloroplasts of *Pisum sativum* L. cv. Alaska by combined gas-chromatography–mass spectrometry. *Planta, Berlin*, **117**, 129–82.

Rivier, L., Milon, H. & Pilet, P. E. (1977). Gas chromatography–mass spectrometric determinations of abscisic acid levels in the cap and the apex of maize roots. *Planta, Berlin*, **134**, 23–7.

Schlenk, H. & Gellerman, J. L. (1960). Esterification of fatty acids with diazomethane on a small scale. *Analytical Chemistry*, **32**, 1412–14.

T. M. WARD, M. WRIGHT, J. A. ROBERTS, R. SELF & DAPHNE J. OSBORNE

Analytical procedures for the assay and identification of ethylene*

Equivalent terms: $mm^3 dm^{-3} = \mu l \ litre^{-1}$; $10^{-12} m^3 dm^{-3} = nl \ litre^{-1}$; $dm^3 = litre$

Introduction

Long before evidence appeared in the scientific literature man knew that air pollutants, smokes and fumes could disturb the normal growth and development of plants. Leaks of illuminating gas (the early town gas made from combusting coal) led to the early work on volatile components causing premature shedding of leaves (Girardin, 1864), stunting, twisting and abnormal horizontal growth of shoots (Molisch, 1884), the fading of flowers (Knight & Crocker, 1913) and premature fruit ripening (Sievers & True, 1912).

Ethylene was identified as the active volatile that caused many of the observed effects (see Crocker, 1948). Eventually, it was established that ethylene is a natural plant product (Gane, 1934), and that the level of ethylene production could be regulated by the level of auxin in the tissue (Morgan & Hall, 1962, 1964). The era of work on ethylene as an important plant hormone had arrived.

Some properties of ethylene ($CH_2{=}CH_2$)

Ethylene is an unsaturated hydrocarbon with molecular weight of 28.05 and boiling point of $-103\ ^\circ C$. It is flammable, colourless and lighter than air (relative density $= 0.978$; air $= 1.0$). At normal pressures (760 mm Hg) the solubility in water at $0\ ^\circ C$ is approximately $315\ mm^3\ dm^{-3}$ and at

---

* All communications regarding this contribution should be addressed to Daphne J. Osborne.

25 °C approaches 140 mm^3 dm^{-3}. At 25 °C the diffusion equilibrium at 1 mm^3 dm^{-3} between air and water is approximately 10000:1.

Methods of assaying ethylene from plants

The levels of ethylene produced by plants are low and of the same order as those of other hormones (usually < 10 ng g^{-1}). For this reason the early workers found considerable difficulties in determining rates of production of the gas. The most sensitive methods then available were bioassays.

Bioassays

The 'triple response' of the etiolated pea plant, developed by Neljubow (1901) is still a useful indicator of the presence of ethylene. Seven-day-old plants enclosed in vessels containing concentrations of 0.01–0.1 mm^3 dm^{-3} ethylene in air, show a measurable reduction in stem elongation within a few hours; at 1 mm^3 dm^{-3} the growth proceeds at a reduced rate and in a horizontal rather than a vertical direction, and within several days a visible swelling of the stem in the elongating region is apparent. Other bioassays with a similar dose–response to ethylene include the downward epinastic movement in tomato leaves (Crocker, Zimmerman & Hitchcock, 1932), acceleration of abscission in debladed bean petioles or cotton cotyledons (Addicott, 1970) and the hook opening reponse of *Phaseolus* seedlings (Kang & Ray, 1969).

Because the test tissue is enclosed for periods of hours or days, all the bioassay methods suffer from modifications of the response by other volatiles, including changes in the levels of CO_2 and oxygen. Age and condition of the test plants are also important.

All the bioassays lack specificity however, since propylene, acetylene and butylene will give similar responses in plants although at 100–1000 times higher concentrations than ethylene.

Now that gas chromatographic facilities are commonly available for identification and quantification of low molecular weight hydrocarbons, bioassays for ethylene are rarely used.

Gas chromatographic assay

Gas–solid chromatography on prepared columns of alumina, Poropak (a preparation of PVC) or silica is by far the most satisfactory and sensitive method for ethylene analysis in use today (Burg & Thimann, 1959). With instruments fitted with flame-ionisation detectors (Pratt & Goeschl, 1969) ethylene can be clearly distinguished from other low molecular weight hydrocarbons, and concentrations as low as 5–10 10^{-12} m^3 dm^{-3} in a 1 cm^3 volume of air can be measured with accuracy within

1–5 min. As many determinations can be performed in a short time, the method is very suitable for biological experimentation.

The machines in use for ethylene assays in the Unit of Developmental Botany, Cambridge, are Pye 104 models with single columns and flame-ionisation detector heads, with standard ionisation amplifiers. Each column oven is operated isothermally and normally, industrial-grade gases are used (British Oxygen Company Ltd, Hadleigh Road, Ipswich, Suffolk) – nitrogen as the carrier gas and hydrogen and air for the detector. It is important that incoming nitrogen and hydrogen are filtered through gas purifying bottles (molecular sieve No. 13 X, Pye Unicam, York St, Cambridge). This precludes the entry to the chromatographic column of particulate contaminants, water vapour and light hydrocarbons from the cylinders. Each chromatograph is linked to a variable speed recorder (Servoscribe 2, Smith Industries Ltd, Waterloo Rd, Cricklewood, London NW2) set at 10 mV.

An integrator can usefully be attached to the chromatograph to quantify large peak areas. However, if the peaks are small, quantification can then be inaccurate. For low concentrations, therefore, determinations are best made directly from the chart scans. Providing the peak height is directly proportional to the amount of ethylene in the injected sample (this can be ascertained against standards) measurement of the peak height alone is satisfactory.

A major decision for gas chromatography of ethylene is the choice of the solid stationary phase in the column. A comparison of the chart-recorded traces for the same gas mixture after separation in four different columns is shown in Fig. 1. Working at near optimal temperatures, silicic acid (Fig. 1a), silica gel (Fig. 1b) and alumina (Fig. 1d) all give short retention times for ethylene, and are therefore highly acceptable for rapid analysis of large numbers of samples. With silicic acid and silica gel (Fig. 1a, b) the ethylene peak is very close to that of ethane. The resolution of these peaks can be critical, particularly when increases in the ethane to ethylene ratio occur. (This situation can be indicative of a reduced oxygen tension around the plant.) Alumina F_1 offers the advantage of the larger difference in retention time between ethane and ethylene. Silicic acid is probably the most suitable solid phase for the higher hydrocarbons, but for good separation of acetylene, Poropak T is satisfactory (Fig. 1c).

Optimising column performance. Individual columns vary considerably in their performance, depending upon the preparation of the solid phase. Thus it is important to determine for each column the optimum operating conditions for maximum resolution with largest response above background (Burchfield & Storrs, 1962).

SILICIC ACID

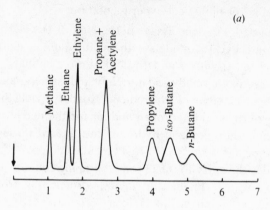

SILICA GEL

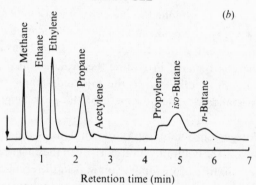

Retention time (min)

Fig. 1. Separation of a gas mixture containing methane, ethane, propane and propylene (each at 50 mm³ dm⁻³), acetylene (10 mm³ dm⁻³), iso-butane and n-butane (each 40 mm³ dm⁻³) and ethylene (100 mm³ dm⁻³). 0.5 cm³ of the gas mixture was injected into the Pye 104 gas chromatograph set at attenuation 200. Chart scans were set at 30 mm min⁻¹ for each. Column characteristics are as follows:

(a) Silicic acid (80–100 mesh) from Mallincrodt, UIA Camlab Chemicals, Trinity Hall Industrial Estate, Nuffield Road, Cambridge, in 152×0.4 cm (internal diameter) glass column with metal seal: nitrogen and hydrogen flow rates 30 cm³ min⁻¹: air (uncritical) 0.5 dm³ min⁻¹: operating temperature 180 °C: column efficiency = 902 theoretical plates m⁻¹.

(b) Silica gel (100–150 mesh) from Hopkins & Williams, P.O. Box 1, Romford, Essex, in 80×0.4 cm (internal diameter) non-standard glass column with glass end: nitrogen and hydrogen flow rates 30 cm³ min⁻¹: air (uncritical) 0.5 dm³ min⁻¹: operating temperature 110 °C: column efficiency = 791 theoretical plates m⁻¹.

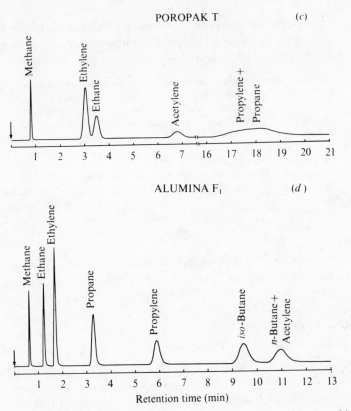

(c) Poropak T (80–100 mesh) from Phase Separations, Deeside
Industrial Estate, Queensferry, Dyfed, UK in 152×0.4 cm (internal
diameter) glass column with metal seal: nitrogen and hydrogen flow
rates 40 cm³ min⁻¹: air (uncritical) 0.5 dm³ min⁻¹: operating
temperature 60 °C: column efficiency = 807 theoretical plates m⁻¹.

(d) Alumina F₁ (80–100 mesh) from J. J. Chromatography Ltd,
Hardwick Trading Estate, Kings Lynn, Norfolk, in 152×0.4 cm
(internal diameter) glass column with metal seal: nitrogen and
hydrogen flow rates 40 cm³ min⁻¹: air (uncritical) 0.5 dm³ min⁻¹:
operating temperature 110 °C: column efficiency = 1657 theoretical
plates m⁻¹.

As the temperature is raised the peak heights are increased, but the
retention times are reduced. The decreased interval in retention time
between the components of a gas mixture can then lead to crowding or
overlapping of the peaks. Increasing the flow rate of the carrier gas results
in a similar sequence of events. For best separation and maximum signal,
each gas has, therefore, specific optimum requirements in each individual
column. Flow rates of hydrogen and air to the detector should be regulated
in relation to that of the carrier gas.

After injection of a series of samples containing high levels of water

Table 1. *The retention times (minutes and seconds) of authentic gases separated on columns under the conditions described in the legend to Fig. 1*

	Silica gel	Silicic acid	Alumina F_1	Poropak T
Methane	0–29	0–36	1–02	0–50
Ethane	0–57	1–12	1–32	3–28
ETHYLENE	1–18	1–40	1–48	3–02
Propane	2–09	3–14	2–36	17–02
Acetylene	2–26	10–50	2–34	6–48
Propylene	4–21	5–49	3–54	18–00
iso-Butane	4–48	9–22	4–25	66–30
n-Butane	5–34	10–54	5–04	91–12

vapour, performance of the column can become impaired. This is readily rectified by reactivating the column for 20–30 min at a higher temperature (approximately +50 °C). When ready-packed columns obtainable from the makers are used, they should either be activated (Poropak T) or conditioned (Alumina F_1) before use as recommended. Silica gel or alumina columns operating at 110 °C with carrier gas flow rates of 30 or 40 cm^3 min^{-1}, have been preferred for ethylene separation in our own work. Poropak T at 60 °C offers no advantages over these, particularly since there is poor resolution at higher temperatues. A summary of retention times for the four columns is set out in Table 1.

Comparisons of minimum detectable ethylene concentrations and the peak height for a given concentration at optimal operating conditions for each of the four columns is shown in Table 2. We can recommend alumina F_1 for detection of the low levels of ethylene encountered from plant tissue (see also Abeles, 1973).

Table 2. *The limits of detection of ethylene on different columns under the conditions described in the legend to Fig. 1*

	Minimum detectable ethylene concentration $(10^{-12} m^3 dm^{-3})$[a]	Peak height on chart recorder (units)[b]
Silica gel	15	34
Silicic acid	12	43
Alumina F_1	7	70
Poropak T	17	30

[a] Signal = 2×noise level.
[b] 0.5 ml of 10^{-12} m^3 dm^{-3} injected. Attenuation 200.

Identification of ethylene by GC–MS

Despite the simple structure in comparison with the other plant hormones, it is not possible to assign an ethylene identity to a column peak using co-chromatography or retention time data from only one column. Co-chromatography on two columns whose separation is based on different properties (e.g. Poropak and silica gel) adds confidence. Conclusive identification can be established by combined gas chromatography and mass spectrometry (GC–MS).

This technique has been used for the identification of ethylene and other volatiles produced by cultures of *Agaricus bisporus* (Turner *et al.*, 1975). For these cultures, and indeed for most plant tissue, collection and concentration of the volatiles is necessary to obtain sufficient material for spectral analysis. The method outlined in 'Concentration by condensation', below, was adopted.

The concentrated volatiles were then separated on a silicic acid column (see legend, Fig. 1*a*), fitted to a Pye 104 gas chromatograph with argon (rather than nitrogen) as the carrier gas. The gas flow emerging from the column was divided in the ratio 2:1 between the silicone membrane separator inlet to the mass spectrometer, and the flame-ionisation detector of the gas chromatograph. Mass spectra were obtained on AE1 MS 902 mass spectrometer and recorded in the range 12–100 m/e using a 10 s decade^{-1} scan speed and an ionisation energy of 70 eV. The volatiles were identified by comparing the mass spectra of the sample components with those for authentic compounds as listed by Cornu & Massot (1966). For the *A. bisporus* cultures it was shown that although the peaks assigned to propane, propylene and *n*-butane contained traces of additional material, those of ethane and ethylene comprised these compounds alone.

Methods for concentrating ethylene for analysis

Because ethylene is volatile and the levels produced by plants are generally so low, a number of methods for trapping, accumulating or concentrating the gas have been employed for obtaining samples that can be committed to gas chromatographic analysis.

Concentration by condensation

This method has advantages when the objective is to concentrate all volatile components including ethylene. The following procedure has been used to condense sufficient quantities of the light hydrocarbons produced by cultures of *A. bisporus* to permit their identification by GC–MS (see previous section). Vapour (1500 cm^3) from the head space of a culture flask (containing 250 g compost) is first dried by passing it, at the rate of 10 cm^3 min^{-1}, through a 46×0.4 cm column packed with 40–60 mesh gelatin (Difco Laboratories, Detroit, Michigan, USA – activated before use by preheating at 90 °C overnight with a purging flow of argon at 40 cm^3 min^{-1}). Up to 2 dm^3 of saturated vapour can be dried through this column without serious loss of any volatile component.

The volatiles are then condensed in a stainless steel concentric tube trap (Swoboda & Lea, 1965), immersed in liquid oxygen (−183 °C) so any argon present is vented to the atmosphere. For analysis of volatiles the trap is heated by immersion in water at 60 °C with simultaneous application of 8 A at 10 V (d.c.) through the metal tube and the volatiles are led directly to the gas chromatograph column.

Continuous flow collection with chemical reactants

Ethylene and other olefins readily form co-ordination products with a variety of organo-metallic compounds in solution, and the reaction with mercuric perchlorate is the basis of one of the most-used methods for quantitative absorption and subsequent quantitative liberation of ethylene from the solution (Young, Pratt & Biale, 1952). The method is very useful when production by a tissue enclosed in a gas flow system is to be monitored over many hours or days and frequent sampling is not required, or when samples must be stored before analysis. It has been adopted, for example, to determine the changes in rates of ethylene production in *Xanthium* leaves at different stages of growth and senescence (Osborne & Hallaway, unpublished).

Three to four grams of leaf tissue are enclosed in 3 dm^3 flasks lined with moist filter paper and sealed with stoppers containing inlet and outlet tubes. Air drawn through the system is first depleted of ethylene by bubbling it at the rate of approximately 2–3 dm^3 h^{-1} through a fine sintered glass head in a Quickfit tube containing 50 cm^3 of 0.25 mol dm^{-3} mercuric perchlorate in 2 mol dm^{-3} perchloric acid (MP) at 0 °C. (For preparation see Young, Pratt & Biale, 1952; Abeles, 1973, p. 21.) It is then bubbled sequentially through sintered heads in two Quickfit flasks each containing 100 cm^3 distilled water at the same temperature (usually 25 °C) as the plant tissue under test. After passage through the tissue flask, the air stream is led through a vessel held at 0 °C to condense excess water vapour, and then bubbled through a sintered glass head in 20 cm^3 of MP in 50 cm^3 Quickfit tubes where the ethylene produced by the plant tissue is trapped. At intervals of several hours these collection tubes are removed and the total contents transferred to volumetric flasks, made up to 25 cm^3 with MP, and stored in glass (not polythene) containers at 2 °C until analysed. These containers must be completely air-tight to preclude further absorption of ethylene from the air, and equivalent samples of the 'ethylene-depleted air' (not passed over the plant tissue) must be collected as blanks. For liberating the ethylene, duplicate 4 cm^3 aliquots are transferred to screw-topped glass containers of measured volume. McCartney bottles with a perforation in the metal top and fitted with a rubber lining are suitable for the purpose. 2 cm^3 of 4 mol dm^{-3} HCl or LiCl are then injected through the perforation and rubber lining into the perchlorate solution to release the ethylene. All vials are vigorously shaken and should be incubated for at least 90 min in a water bath at 25 °C before analysis. Blank vials, and a set containing MP and with known ethylene contents, are treated similarly. 1 cm^3 samples of head space containing the liberated ethylene are withdrawn from the vials with gas-tight syringes and injected directly into a gas chromatograph. Recovery is 99% for ethylene production levels not in excess of 4000 10^{-12} m^3 h^{-1}. Most tissues do not exceed 0.4 10^{-12} m^3 g^{-1} h^{-1} but ripe fruits and senescent leaves may produce as much as 4–10 10^{-12} m^3 g^{-1} h^{-1}.

Ethylene accumulation in the vapour phase

In this method, which is most convenient for measuring ethylene produced from amounts of material as small as 1 g, the tissue is enclosed in a container fitted with a gas sampling port e.g. a rubber Subaseal (W. Freeman Ltd, Barnsley, Yorks.). The ethylene that is produced diffuses from the tissue into the small volume of surrounding air and accumulates. Gas samples can then be withdrawn at intervals, through the Subaseal, with a gas-tight syringe (the disposable plastic type, Gillette Surgical, Great West Road, Isleworth, Middlesex, is suitable for this purpose), and analysed directly by gas chromatography.

The technique must, however, be used with some degree of caution.

Enclosure may modify the physiological processes under observation. In abscission experiments, for instance, enclosure will induce premature abscission as a result of maintaining the tissue in ethylene of a higher concentration than that of non-enclosed samples (Jackson & Osborne, 1970).

Oxygen may become limiting in small-volume containers, and the tissues then respire anaerobically, affecting both the ethylene-producing system (which is aerobic) and the physiological process under observation. An indication of anaerobiosis is reflected in an increase in the ratio of ethane to ethylene production. With this in mind, care must be taken to ensure that frequent aerations of the enclosing vessel are carried out, and time course analyses are advised.

It is essential that all aerations are performed with air that is not contaminated by any additional ethylene above the normal background. An airline of outdoor air can be pumped in for this purpose, but must be routinely checked for ethylene contamination. Ethylene levels in rural areas are usually less than 5 ppb, but may be considerably higher in urban areas (Abeles, 1973). The levels are barely detectable above background in scans of air samples taken at the Unit in Cambridge. The possibility of contaminating ethylene from bonfires (as high as 4 ppm) or car exhausts (up to 400 ppm), or from aerosols or polishes, must be carefully noted and guarded against. If possible, all ethylene monitoring experiments should be carried out in a laboratory designed only for such work.

Ethylene is readily absorbed by many materials used in experiments. New subaseals, disposable hypodermics or any plastic material used in an enclosure system, should be left open to the air for at least a day before use. In all enclosure systems it is important that the empty flask, with its subseal, should be monitored for each experiment to ensure that ethylene liberation is not attributable to the rubber closure.

If sterile conditions are required, Subaseals should not be autoclaved or irradiated with UV light. Sterilisation with absolute ethanol at room temperature is satisfactory, and does not lead to a liberation of ethylene from the rubber.

It is possible to determine ethylene production by 0.5 g of tissue in a 5 cm^3 enclosure with sampling times as short as 30 min. Temperature must, however, be kept constant during the experiment and during gas sampling to preclude errors caused by volume changes.

Precautions to note

Determinations of ethylene accumulated during one enclosure period only provide an average value that does not necessarily represent production by the tissue under test. With mushroom compost, for example, considerable desorption of ethylene produced before enclosure may initially occur from the medium, and this can obscure a measurement of the rate of production of newly synthesised ethylene. Time-course determinations are therefore important, and provided the tissue is not undergoing physiological change, a linear value for the rate can usually be established in a few hours.

Another feature to be noted is the so-called 'wound ethylene' production by plant tissues. For most plants, this several-fold increase commences within 30–45 min whenever a plant is touched or handled in any way, and can persist for several hours (Jackson & Osborne, 1970; Irvine & Osborne, 1973). Measurements of ethylene production of a tissue taken after 30–45 min of placing in the enclosure vessel will therefore include the wound ethylene component in addition to the basal rate. For basal rates of wound ethylene production, therefore, determinations are usually made either before 30 min or after the wound ethylene production has subsided. The latter can be determined by closing the plant container and sampling the gas phase at suitable intervals; when the rate of ethylene production has declined to a steady level, it is then appropriate to determine the basal rate of production of the tissue. Modifications of the basal ethylene production rate in the absence of wound ethylene, i.e. in response to chemical or other treatments, can be assayed in this way. Jackson & Campbell (1976) have, however, suggested that ethylene determinations made during the first 20 or 30 min, and before the wound-induced rise, may be a better estimate of basal production rates of a tissue than those made after the wound ethylene production has subsided.

It should be borne in mind that none of the methods described in this section give a direct value for the concentration of ethylene within a plant tissue. They provide a value for the rate at which the ethylene produced equilibrates between the tissue and the enclosing air. Whereas such comparisons between similar types of tissue may be valid, comparisons made between very different types of tissue, e.g. fruits and leaves, are open to criticism as they are subject to the limitations imposed by different rates of equilibration between epidermis, cuticle and cut surfaces of the two types of tissue and to differences in their ratio of surface area to volume (see Abeles, 1973, pp. 37–41; Burg & Burg, 1962).

Extraction of ethylene from within the plant tissues

Where sizeable cavities exist within the plant organs, i.e. in the cores of apples or the centre of melons, air samples can be withdrawn by hypodermic syringe and injected directly into the gas chromatograph (Lyons, McGlasson & Pratt, 1962). Such assays provide values for the internal air spaces but not for the ethylene dissolved in the internal aqueous phase.

Estimations of total ethylene within a plant can be obtained by subjecting the tissues to vacuum below the surface of a solution in which ethylene is poorly soluble. The liberated gases are collected for subsequent analysis by gas chromatography. The method devised by Beyer & Morgan (1970) is simple to execute. A desiccator is filled with a saturated solution of $(NH_4)_2SO_4$, and a funnel inverted beneath the surface; the end of the funnel is then sealed with a vaccine cap. The plant material is introduced beneath the funnel, the desiccator is closed and vacuum is applied (100 mm Hg for 2 min). Gases present in intracellular air spaces or in solution within the tissue are liberated and accumulate below the vaccine cap. On releasing the vacuum, samples of this gas phase are removed with a syringe and injected into the gas chromatograph.

The values obtained by direct sampling of the internal atmospheres of fruits are less than those obtained from the vacuum extraction method, suggesting that much of the ethylene produced normally remains in solution within the tissue. However, what proportion represents ethylene in solution in the cytoplasm as distinct from cell walls or other free spaces cannot be evaluated. There is a further possibility that vacuum treatment could cause a considerable level of wound ethylene production despite the short period of manipulation, and Beyer & Morgan have shown that lowering the vacuum below 100 mm Hg can increase the levels of ethylene obtained. As the tissue becomes waterlogged, this method does not permit more than one assay from each sample of tissue.

Ethylene production by submerged tissues

Ethylene production is an aerobic process in plants and is generally reduced or inhibited if the tissue is submerged (apple tissue slices, Lieberman, Kunishi, Mapson & Wardale, 1966; Burg & Clagett, 1967). However, certain aquatic plants possessing extensive aerenchyma tissue continue to synthesise ethylene in the light whilst below the water surface. An indirect measure of the rates of production over extended periods can be obtained by the sampling procedure adopted by Musgrave, Jackson & Ling (1972), which is a modification of that described by Beyer & Morgan (1970). The plant tissue is submerged beneath an inverted funnel, the stem of which is capped with a rubber teat. The gas bubbles liberated during photosynthesis contain ethylene and accumulate in the stem of the funnel. Gas samples can be withdrawn through the teat and injected into the gas chromatograph. The method provides an underestimate of production, for considerable losses (20–50%) result from the solubility of ethylene in water.

Assay of radioactive ethylene

Quantitative monitoring of the production of radioactive ethylene from radioactive precursors can most conveniently be carried out by gas chromatography. Any radioactive CO_2 produced can be separated from ethane and ethylene on a Poropak Q column fitted with a thermoconductivity detector and the effluent gases led to a continuous flow radioactive monitoring unit for measurement of radioactivity of the separated fractions (Baur & Yang, 1969).

In the absence of facilities for simultaneous quantification and radioassay, the production of radioactive ethylene from labelled precursors can be assessed by conventional liquid scintillation counting procedures. The following modification of the methods of Yang, Ku & Pratt (1967) has been used successfully in our laboratories for leaf discs, with simultaneous determination of radioactive CO_2 production.

The incubation of tissue with precursor is carried out in 6×2 cm glass vials with a single side arm (Fig. 2) (T. W. Wingent Ltd, Milton, Cambridge, UK). A circle of filter paper moistened with distilled water, or a thin layer of 2% agar is included at the base to maintain a humid atmosphere. Drops (1 mm^3; 50–55 μCi cm^{-3}) of 1-[U-^{14}C]methionine (specific activity 280 mCi mmol^{-1} from Radiochemical Centre, Amersham) are applied on a glass coverslip and a leaf disc (4 mm diameter) placed abaxial surface downwards on each drop. The coverslip is lowered into the vial and a frill of filter paper, moistened with 0.1 cm^3 of 1 mol dm^{-3} KOH is inserted into the side arm to absorb CO_2. A Subaseal, with a suspended glass bucket, is inserted into the vial and the whole incubated for the appropriate period. A sample of the gas (0.5 cm^3) is then withdrawn and the ethylene concentration determined by gas chromatography. Mercuric perchlorate (0.25 mol dm^{-3} in 2 mol dm^{-3} perchloric acid) is then injected into the suspended glass bucket to absorb the remaining ethylene

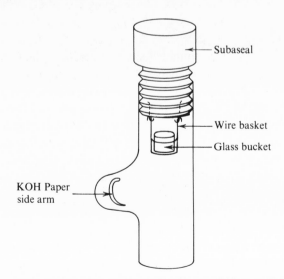

Fig. 2. Reaction flask suitable for use in determining the production of labelled ethylene from a radioactive precursor.

in the vial. After a short interval (30 min) the glass bucket can be removed, attached to a new Subaseal and inserted into a scintillation vial. The KOH paper is removed to another scintillation vial and dried. Measurement of ethylene and CO_2 production from the same leaf discs can be repeated by including new mercuric perchlorate solutions and fresh KOH papers.

Mercuric perchlorate causes precipitation in most scintillation fluids, so the absorbed ethylene has first to be liberated and reabsorbed into a solution that is scintillant-tolerant. This is done by injecting $0.1 \, \text{cm}^3$ of 6 mol dm^{-3} HCl into the glass bucket and incubating for 3 h at room temperature. The ethylene is then reabsorbed into $0.3 \, \text{cm}^3$ mercuric acetate (0.1 mol dm^{-3} in methanol) injected into the base of the vial, which is then incubated and shaken for at least 1 h at 4 °C. On removal of the Subaseal, $10 \, \text{cm}^3$ of water tolerant scintillation fluid (NE 260, Nuclear Enterprises, Sighthill, Edinburgh) are then added to the mercuric acetate and the mixture counted in the usual way.

Assays show that a high recovery (98–100%) of radioactive ethylene can be achieved by this method.

The dry KOH papers can be counted directly in non-water-tolerant scintillants for determination of the $^{14}CO_2$ production.

Incubation of the [^{14}C]methionine on coverslips in vials in the absence of plant tissue does not produce radioactive ethylene, but can result in some decomposition to liberate CO_2. It is important, therefore, that precursor blanks are included in the experiments.

Conclusion

Because ethylene is a simple volatile molecule which can be readily separated from other volatiles it is by far the easiest of the plant hormones to quantify, once a sample is obtained for analysis.

But the problems of obtaining that sample, and the assurance that can be placed upon the validity of the value so obtained are as acute as those for any of the other plant hormones. Tissues in which active cell division or expansion are taking place, senescing (or ripening) tissue, or tissues that are 'wounded', produce clearly measurable amounts of the gas. Mature or non-growing tissues may evolve so little that it has been questioned whether *in vivo* and undisturbed they produce the gas at all.

The problems of extracting ethylene from plant material are fraught with difficulty, for ethylene exists within the tissue as a gas in the intercellular spaces, and in solution within both the cytoplasm and the free space. The level within the tissue is in part a function of the rate of synthesis, but is regulated also by the solubility and rates of physical diffusion in aqueous and gaseous phases.

The values for rates of ethylene production following any of the extraction or concentration procedures described are therefore a reflection of the extent to which ethylene has been recovered from the different compartments. Values for total ethylene contents suffer the same objections, and include possible overestimates resulting from wound ethylene. Unlike the other hormones, ethylene appears to be the terminal product of a metabolic reaction chain, and, as Beyer (1975) has shown for pea plants, less than 1% of applied ethylene is converted to other products. At least therefore, serious underestimates of ethylene values resulting from ethylene degradation seem unlikely.

Note added in proof
Evidence (Jerie, P. H. & Hall, M. A. (1978). *Proceedings of the Royal Society of London, Series B*, in press) indicates that cotyledons of developing seeds readily convert [^{14}C]ethylene to [^{14}C]ethylene oxide.

References
Abeles, F. B. (1973). *Ethylene in Plant Biology.* New York and London: Academic Press.

Addicott, F. T. (1970). Plant hormones in the control of abscission. *Biological Reviews*, **45**, 485–524.

Baur, A. & Yang, S. F. (1969). Ethylene production from propanal. *Plant Physiology*, **44**, 189–202.

Beyer, E. M. (1975). C_2H_4: Its incorporation and metabolism by pea seedlings under aseptic conditions. *Plant Physiology*, **56**, 273–8.

Beyer, E. M. & Morgan, P. W. (1970). A method of determining the concentration of ethylene in the gas phase of vegetative plant tissues. *Plant Physiology*, **46**, 352–4.

Burchfield, H. P. & Storrs, E. E. (1962). *Biochemical Applications of Gas Chromatography*, pp. 77–88, 105–13. New York and London: Academic Press.

Burg, S. P. & Burg, E. A. (1962). Role of ethylene in fruit ripening. *Plant Physiology*, **37**, 179–89.

Burg. S. P. & Clagett, C. O. (1967). Conversion of methionine to ethylene in vegetative tissue and fruits. *Biochemical and Biophysical Research Communications*, **27**, 125–30.

Burg, S. P. & Thimann, K. V. (1959). The physiology of ethylene formation in apples. *Proceedings of the National Academy of Sciences, USA*, **45**, 335–44.

Cornu, A. & Massot, R. (1966). *Compilation of Mass Spectral Data.* London: Heyden.

Crocker, W. (1948). Physiological effects of ethylene and other unsaturated carbon-containing gases. In *Growth of Plants – Twenty Years' Research at Boyce Thompson Institute*, pp. 139–71. New York: Reinhold.

Crocker, W., Zimmerman, P. W. & Hitchcock, A. E. (1932). Ethylene-induced epinasty of leaves and the relation of gravity to it. *Contributions of the Boyce Thompson Institute*, **4**, 177–218.

Gane, R. (1934). Production of ethylene by some ripening fruits. *Nature, London*, **134**, 1008.

Girardin, J. P. L. (1864). Einfluss des Leuchtgases auf die Promenaden und Strassenbaume. *Jahresbericht über die Fortschritte auf dem gesamtgebiete der Agrikultur Chemie, Versuchssta, Berlin*, **7**, 199–200.

Irvine, R. F. & Osborne, D. J. (1973). The effect of ethylene on [1-^{14}C]glycerol incorporation into phospholipids of etiolated pea stems. *Biochemical Journal*, **136**, 1133–5.

Jackson, M. B. & Campbell, D. J. (1976). Production of ethylene by excised segments of plant tissue prior to the effect of wounding. *Planta, Berlin*, **129**, 273–4.

Jackson, M. B. & Osborne, D. J. (1970). Ethylene, the natural regulator of leaf abscission. *Nature, London*, **225**, 1019–22.

Kang, B. G. & Rat, P. M. (1969). Ethylene and carbon dioxide as mediators in the response of the bean hypocotyl hook to light and auxins. *Planta, Berlin*, **87**, 206–16.

Knight, L. I. & Crocker, W. C. (1913). Toxicity of smoke. *Botanical Gazette, Chicago*, **55**, 337–71.

Lieberman, M., Kunishi, A., Mapson, L. W. & Wardale, D. A. (1966). Stimulation of ethylene production in apple tissue slices by methionine. *Plant Physiology*, **41**, 376–82.

Lyons, J. M., McGlasson, W. B. & Pratt, H. K. (1962). Ethylene production, respiration and internal gas concentrations in cantaloupe fruits at various stages of maturity. *Plant Physiology*, **37**, 31–6.

Molisch, H. (1884). Sitzungsberitche der Kaiserl. *Akademie der Wissenschaften* (Wein), **90**, 111–96.

Morgan, P. W. & Hall, W. C. (1962). Effect of 2,4-dichlorophenoxyacetic acid on the production of ethylene by cotton and grain sorghum. *Physiologia Plantarum*, **15**, 420–7.

Morgan, P. W. & Hall, W. C. (1964). Accelerated release of ethylene by cotton following applications of indolyl-3-acetic acid. *Nature, London*, **204**, 99.

Musgrave, A., Jackson, M. B. & Ling, E. (1972). *Callitriche* stem elongation is controlled by ethylene and gibberellin. *Nature New Biology*, **238**, 93–6.

Neljubow, D. N. (1901). Über die Horizontale Nutation der Stengel von *Pisum sativum* und Einiger Anderen. *Pflanzen Beiträge und Botanik Zentralblatt*, **10**, 128–139.

Pratt, H. K. & Goeschl, J. D. (1969). Physiological roles of ethylene in plants. *Annual Review of Plant Physiology*, **20**, 541–84.

Sievers, A. F. & True, R. H. (1912). A preliminary study of the forced curing of lemons as practised in California. *US Department of Agriculture and Bureau of Plants 2nd Bulletin*, 232.

Swoboda, P. A. T. & Lea, C. H. (1965). The flavour volatiles of fats and fat containing foods. Part 11. A gas chromatographic investigation of volatile autoxidation products from sunflower oil. *Journal of the Science of Food and Agriculture*, **16**, 680–9.

Turner, E. M., Wright, M., Ward, T. M., Osborne, D. J. & Self, R. (1975). Production of ethylene and other volatiles and changes in cellulase and laccase activities during the life cycle of the cultivated mushroom, *Agaricus bisporus*. *Journal of General Microbiology*, **91**, 167–76.

Yang, S. F., Ku, H. S. & Pratt, H. K. (1967). Photochemical production of ethylene from methionine and its analogues in the presence of flavin mononucleotide. *Journal of Biological Chemistry*, **242**, 5274–80.

Young, R. E., Pratt, H. K. & Biale, J. B. (1952). Manometric determination of low concentrations of ethylene. *Analytical chemistry*, **24**, 551–5.

Index

abscisic acid (ABA)
   identification of: by electron capture,
      124–6; by GC–MS, 126–7; by GLC of
      methyl ester and of 2-*trans* isomer,
      120–4; by optical activity, 118–19
   not synonymous with inhibitor β, 115–17
   preparation of methyl ester of, 119–20
   purification of, from plant extracts,
      127–31; by HPLC, 8–9
   quantitative analysis of, 131–3; by mass
      fragmentography, 74
acetone as solvent, drawbacks of, 82
acetonides of dihydroxylated GAs, for
      GC, 86
n-alkanes, from Parafilm, 80
AMP, radioactive: recovery of, from plant
      extracts by different methods, 99
anaerobiosis of plant tissues, and ratio of
      ethane to ethylene produced, 143
aquatic plants, production of ethylene by,
      146
auxin, use of term, 2

benzyladenine, radioactive: recovery of
      metabolites of, from plant extracts by
      different methods, 99, 100
benzyl esters of gibberellins (GABEs)
   N,N'-dimethylformamide dibenzyl acetal
      reagent for preparing, 56–7;
      esterification procedure with, 58–9
   HPLC (analytical) of, 59, 63; effect of
      numbers of benzyl and hydroxyl
      groups on, 60
   mass spectra of, 56, 64–8
bioassays
   of ABA and inhibitor β, 115, 116, 117
   of ethylene, lack specificity, 136
   of GAs, 73–4, 79
borohydride, sodium: to confirm identity
      of ABA in GC, 122
n-butylboronate derivatives of GAs, for
      GC, 86

cellulose phosphate, as cation exchange
      material for cytokinins, 101

chromatography, *see* column, gas,
      gas–liquid, gas–solid, gel permeation,
      high performance liquid, liquid, paper,
      and thin-layer chromatography
column chromatography
   on charcoal, of semi-purified GAs, 43,
      46–7, 73
   of cytokinins, 102, 104–6: pressurised
      system for, 106, 107
   on DEAE cellulose and then PVP, in
      separation of IAA, 7–8
   on PVP, not recommended for
      cytokinins, 106, or for GAs, 47
   on PVP, to remove light-scattering
      compound from
      2-methylindole-α-pyrone, 30
columns
   optimising performance of, for GSC of
      ethylene, 137, 139–40
   wide-bore for preparative and narrow
      for analytical HPLC, 42, 56
computerised signal processing, for MS,
      20, 71, 88–9, 90
condensation, of volatiles containing
      ethylene, 141–2
cytokinins
   analytical methods for, 97–8; column
      chromatography, 102, 104–6, 107;
      gas–liquid chromatography, 108–10;
      ion exchange, 100–1; paper and
      thin-layer chromatography, 106, 108
   extraction of, 98–100, and solvent
      partition, 102, 103
   methods of identifying, 110–12
   prospects for use of HPLC for, 110,
      112

diffusion into agar, separation of IAA by, 4
diazoethane, for methylation of ABA,
      119–20, of GAs, 84, and of IAA, 10
   potassium hydroxide as contaminant of,
      84, 85
dimethylsulphoxide, as modifier in
      analytical HPLC of GABEs, 61, 62,
      63

dioxane, as solvent in esterification of
GAs to GABEs, 58
drying
in purification of GAs, 44–5, 82
of TMSi derivatives of cytokinins, 109,
110

Ehrlich's reagent, for detection of IAA in
TLC, 9–10
electron capture detector
for ABA methyl ester, 124–6, 131
for IAA, 10, 11
emulsions formed in purification
processes, methods of removing, 45,
129
enzymatic methods, for identifying
cytokinins, 112
ethane
anaerobiosis, and ratio of ethylene to,
143
resolution of ethylene and, in GSC, 137
ethylene, 135
assays of: bioassays, 136; GSC, 136–40
assays of radioactive, 146–8
concentration of, for analysis: by
accumulation in vapour phase, 143; by
condensation, 141–2; by continuous
flow collection with mercuric
perchlorate, 142–3; sources of
contamination, 144
extraction of, from within plant tissues,
145–6
identification of, by GC–MS, 141
production of: by aquatic plants, 146; on
wounding or handling of plants,
144–5, 148, 149
as terminal product, 149
extraction from plant tissues
of cytokinins, 98–100
of ethylene, by vacuum, 145–6
of GAs, 43–5, 82
of IAA: from fresh tissues, 4–5; from
tissues frozen in liquid nitrogen, 29

fatty acids: non-polar, as contaminants in
GC of GAs, 81
flame ionisation detector
for ABA, 122–3, 124, 125, 131
for ethylene, 136, 137, 141
for IAA, 10, 11, 13
fluorescence, in spectrophotofluorimetry of
IAA derivative
background, 31, 32, 33, 35
effect on: of catalyst used in preparing
derivative, 33; of plant extracts, 34,
35
rate of decay of, 31
formic acid, for stationary phase in
preparative HPLC, 48

gas chromatography (GC), of GA
derivatives, 86–7
gas chromatography–mass spectrometry
(GC–MS), 11–12
of ABA methyl ester, 126–7
of cytokinin TMSi derivatives, 111–12
of GA methyl and methyl TMSi
derivatives, 41, 79–80, 88–93;
reference spectra for, 93
of IAA, 13–16; quantitative
determination, 16; worked example,
19–20
identification of ethylene by, 141
gas–liquid chromatography (GLC), 10
of ABA methyl ester, 119–21, 133
of cytokinin TMSi derivatives, 108–10
of IAA derivatives, 10–11, 13
gas–solid chromatography (GSC), for
ethylene analysis, 136–7, 138
optimising column performance in, 137,
139–40
gel permeation chromatography (GPC), of
extracts containing GAs, 43, 45–6,
73
Gibberella fujikuroi, high contents of GAs
in, 42
gibberellins (GAs)
analysis of endogenous: qualitative,
72–3; quantitative, 73–4
conjugated: hydrolysis of, 83–4; in
purification, 43, 44
deuterium-labelled, as internal markers,
74
double-bond isomers of, not separated
in preparative HPLC, 56
effect of number of hydroxyl groups in:
on esterification to benzyl esters, 58;
on HPLC, 53–4, 55
extraction of, and solvent partitioning,
43–5; group separatory procedures,
45–7
free, in purification, 44, 46
GC–MS of methyl and methyl TMSi
derivatives of, 41, 79–80, 88–93;
preparation of samples for,
80–3
HPLC of (preparative), 42, 47–55
HPLC of benzyl esters of (analytical),
59–65, 83
HPLC–MS of, 41–3
inhibitors of, in bioassay, 73, 74, 79
mass fragmentography of, 74
MS of benzyl esters of, 56, 64–8; from
seedlings after treatment with
radioactive GA9, 68–71
paper chromatography of, 117
preparation of benzyl esters of, 56–8,
and separation by preparative HPLC,
58–9